Exam Success

IEE Wiring Regulations 2382-10

Paul Cook and Jonathan Elliott

Reprinted 2008

ISBN: 978-0-86341-885-3

First edition published 2006 as *Exam Success:
IEE Wiring Regulations 2381*

Cover and book design by CDT Design Ltd
Typeset in Congress Sans and Gotham
Printed in the UK by Ashford Colour Press

With thanks to Brian Scaddan

Exam Success
IEE Wiring Regulations 2382-10
Paul Cook and Jonathan Elliott

City & Guilds Level 3 Certificate in the Requirements for Electrical Installations (BS 7671: 2008)

City & Guilds is the UK's leading provider of vocational qualifications, offering over 500 awards across a wide range of industries, and progressing from entry level to the highest levels of professional achievement. With over 8500 centres in 100 countries, City & Guilds is recognised by employers worldwide for providing qualifications that offer proof of the skills they need to get the job done.

Copies may be obtained from:
Teaching & Learning Materials
City & Guilds
1 Giltspur Street
London EC1A 9DD
For publications enquiries:
T +44 (0)20 7294 4113
F +44 (0)20 7294 3414
Email learningmaterials@cityandguilds.com

The Institution of Engineering and Technology is the new institution formed by the joining together of the IEE (The Institution of Electrical Engineers) and the IIE (The Institution of Incorporated Engineers). The new institution is the inheritor of the IEE brand and all its products and services including the IEE Wiring Regulations (BS 7671) and supporting material.

Copies may be obtained from:
The Institution of Engineering and Technology
P.O. Box 96
Stevenage
SG1 2SD, UK
T +44 (0)1438 767 328
Email sales@theiet.org
www.theiet.org

Contents

Introduction

How to use this book

This book has been written as a study aid for the City & Guilds Level 3 Certificate in the Requirements for Electrical Installations (BS 7671: 2008) (2382-10). It sets out methods of studying, offers advice on exam preparation and provides details of the scope and structure of the examination, alongside sample questions with fully worked-through answers. Used as a study guide for exam preparation and practice, it will help you to reinforce and test your existing knowledge, and will give you guidelines and advice about sitting the exam. You should try to answer the sample test questions under exam conditions (or as close as you can get) and then review all of your answers. This will help you to become familiar with the types of question that might be asked in the exam and also give you an idea of how to pace yourself in order to complete all questions comfortably within the time limit. This book cannot guarantee a positive exam result, but it can play an important role in your overall revision programme, enabling you to focus your preparation and approach the exam with confidence.

IEE Wiring Regulations Seventeenth Edition

You will need a copy of the *IEE Wiring Regulations Seventeenth Edition* to be able to answer the sample questions and in order to revise for the examination. The *IEE Wiring Regulations Seventeenth Edition*, also called *BS 7671: 2008 Requirements for Electrical Installations*, is the national standard for the electrical industry in respect of safe use and operation of electrical equipment and systems in the United Kingdom, in accordance with the recommendations of the Institution of Engineering and Technology (IET). It contains rules for the design and erection of electrical installations, so as to provide for safety and proper functioning for the intended use.

City & Guilds Level 3 Certificate in the Requirements for Electrical Installations (BS 7671: 2008) (2382-10)

If you are a practising electrician with relevant experience, you need to achieve this Level 3 Certificate to show you have a sound working knowledge of the format, content and application of the current edition of the wiring regulations, the *IEE Wiring Regulations Seventeenth Edition*. This award is also useful for electricians who have studied electrical installation practice in other countries, and are familiar with either the International Standards in the IEC 60364 Electrical Installations of Buildings series, or European Standard Series HD384 Electrical Installations of Buildings.

This qualification includes all new material and updates to the *IEE Wiring Regulations*. If you already have a certificate for the Wiring Regulations prior to the Seventeenth Edition, 2008, you will need to obtain an update award.

The syllabus covers topics such as protection for safety, selection and erection of equipment, and inspection and testing to meet the standards of the *IEE Wiring Regulations*. If successful, you could go on to take other City & Guilds electrical qualifications, such as Inspection, Testing and Certification of Electrical Installations (2391-10), and In-service Inspection and Testing of Electrical Equipment (2377).

Finding a centre

In order to take the exam, you must register at an approved City & Guilds centre. You can find your nearest centre by looking up the qualification number 2382-10 on www.cityandguilds.com. The IET is an accredited centre and runs online exams in different parts of the country. For more details, see www.theiet.org.

At each centre, the Local Examinations Secretary will enter you for the award, collect your fees, arrange for your assessment to take place and correspond with City & Guilds on your behalf. The Local Examinations Secretary also receives all of your certificates and correspondence from City & Guilds. Most centres will require you to attend a course of learning before entering you for the examination. These can vary between week-long intensive courses and courses running across a term, once a week for two to three hours.

Awarding and reporting

When you complete the City & Guilds 2382-10 Certificate online examination, you will be given your provisional results, as well as a breakdown of your performance in the various areas of the examination. This is a useful diagnostic tool if you fail the exam, as it enables you to identify your individual strengths and weaknesses across the different topics.

A Certificate is issued automatically when you have been successful in the assessment, but it will not indicate a grade or percentage pass. Your centre will receive your Notification of Candidate's Results and Certificate. Any correspondence is conducted through the centre. The centre will also receive consolidated results lists detailing the performance of all candidates entered. If you have particular requirements that will affect your ability to attend and take the examination, then your centre should refer to City & Guilds policy document 'Access to Assessment: Candidates with Particular Requirements'.

Notes

The exam

The exam

The exam

The examination has a multiple-choice format, with 60 questions, which you will have two hours to answer. The test is offered on GOLA, a simple online service that does not require strong IT skills. GOLA uses a bank of questions set and approved by City & Guilds. Each candidate receives randomised questions, so no two candidates will sit exactly the same test.

The exam always follows a set structure, based on the seven parts of the *IEE Wiring Regulations* plus the appendices. The table below outlines the sections of the exam and the number of questions in each section. It also shows the weighting – so you can see how important each section is in determining your final score.

Section	Topic		% weighting	No of questions
1	Part 1	Scope, object and fundamental principles	7	4
2	Part 2	Definitions	5	3
3	Part 3	Assessment of general characteristics	8	5
4	Part 4	Protection for safety	23	14
5	Part 5	Selection and erection of equipment	25	15
6	Part 6	Inspection and testing	10	6
7	Part 7	Special installations or locations	17	10
8	Use of appendices		5	3
	Total		100	60

Sitting a City & Guilds online examination

The test will be taken under usual exam conditions. You will not be able to refer to any materials or publications other than the *IEE Wiring Regulations Seventeenth Edition*. You will not be allowed to take your mobile phone into the exam room and you cannot leave the exam room unless you are accompanied by one of the test invigilators. If you leave the exam room unaccompanied before the end of the test period, you may not be allowed to come back into the exam.

When you take a City & Guilds test online, you can go through a tutorial to familiarise yourself with the online procedures. When you are logged on to take the exam, the first screen will give you the chance to go into a tutorial. The tutorial shows how the exam will be presented and how to get help, how to move between different screens, and how to mark questions that you want to return to later.

Please work through the tutorial before you start your examination.

This will show you how to answer questions and use the menu options to help you complete the examination.

Please note that examination conditions now apply.

The time allowed for the tutorial is 10 minutes.

Click on Continue to start the tutorial or Skip to go straight to the examination.

| Skip | Continue |

Notes

The sample questions in the tutorial are unrelated to the exam you are taking. The tutorial will take 10 minutes, and is not included in the test time. The test will only start once you have completed or skipped the tutorial. A screen will appear that gives the exam information (the time, number of questions and name of the exam).

City& Guilds

Examination: 2382-10 Requirements for Electrical Installations

Number of questions: 60

Time allowed: 120 minutes

Note: Examination conditions now apply.

The next screen that will appear is the Help screen, which will give you instructions on how to navigate through this examination. Please click OK to view the Help screen.

The time allowed for the examination will start after you have left the Help screen.

A warning message will appear 5 minutes before the end of the examination.

OK

After clicking 'OK', the Help screen will appear. Clicking the 'Help' button on the tool bar at any time during the exam will recall this screen.

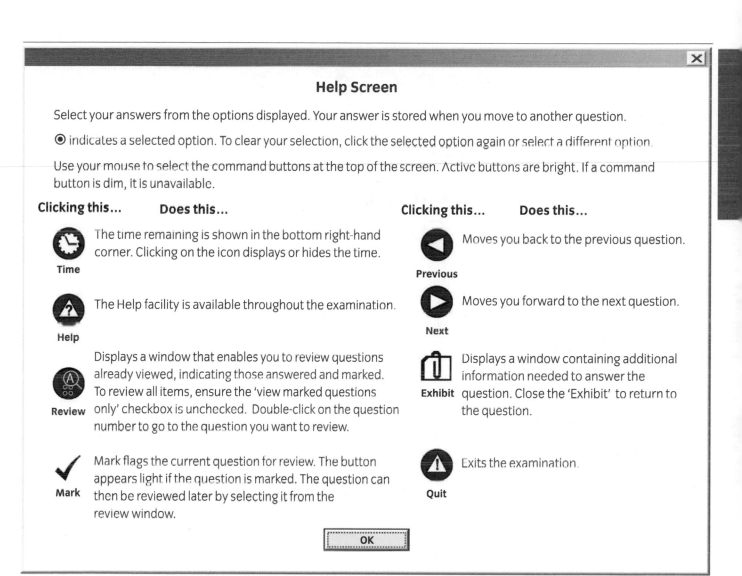

After clicking 'OK' while in the Help screen, the exam timer will start and you will see the first question. The question number is always shown in the lower left-hand corner of the screen. If you answer a question but wish to return to it later, then you can click the 'Flag' button. When you get to the end of the test, you can choose to review these flagged questions.

Notes

BS 7671 applies to

○ a   equipment on board ships

○ b   lightning protection of buildings

○ c   lift installations

○ d   prefabricated buildings.

Question Number 1

If you select 'Quit' on the tool bar at any point, you will be given the choice of ending the test. **If you select 'Yes', you will not be able to go back to your test.**

If you click 'Time' on the tool bar at any point, the time that you have left will appear in the bottom right-hand corner. When the exam timer counts down to five minutes remaining, a warning will flash on to the screen.

Some of the questions in the test may be accompanied by pictures. The question will tell you whether you will need to click on the 'Exhibit' button to view an image.

When you reach the final question and click 'Next', you will reach a screen that allows you to 'Review your answers' or 'Continue' to end the test. You can review all of your answers or only the ones you have flagged. To review all your answers, make sure that the 'view marked questions only' checkbox is unchecked (click to uncheck). After you have completed your review, you can click 'Continue' to end the test.

City&Guilds

You have answered 60 questions out of a total of 60

To check your answers and return to the examination, click on the Review button. If your time has expired, you cannot return to the examination.

If you wish to submit your answers and end the examination, please click the Continue button.

Clicking Continue will end the examination

| Review | Continue |

Once you choose to end the exam by clicking 'Continue', the 'Test completed' screen will appear. Click on 'OK' to end the exam.

Notes

At the end of the exam, you will be given an 'Examination Score Report'. This gives a provisional grade (pass or fail) and breakdown of score by section. This shows your performance in a bar chart and in percentage terms, which allows you to assess your own strengths and weaknesses. If you did not pass, it gives valuable feedback on which areas of the course you should revise before re-sitting the exam.

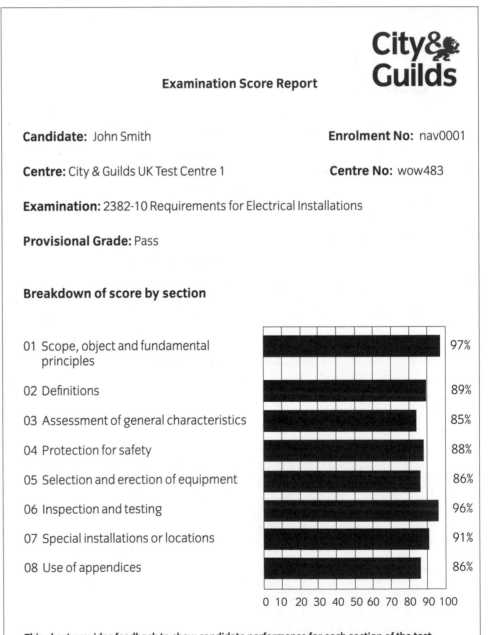

City& Guilds

Examination Score Report

Candidate: John Smith   **Enrolment No:** nav0001

Centre: City & Guilds UK Test Centre 1   **Centre No:** wow483

Examination: 2382-10 Requirements for Electrical Installations

Provisional Grade: Pass

Breakdown of score by section

Section	Score
01 Scope, object and fundamental principles	97%
02 Definitions	89%
03 Assessment of general characteristics	85%
04 Protection for safety	88%
05 Selection and erection of equipment	86%
06 Inspection and testing	96%
07 Special installations or locations	91%
08 Use of appendices	86%

0 10 20 30 40 50 60 70 80 90 100

This chart provides feedback to show candidate performance for each section of the test.
It should be used along with the Test Specification, which can be found in the Scheme Handbook.

Frequently asked questions

When can I sit the paper?
You can sit the exam at any time, as there are no set exam dates. You may need to check with your centre when it is able to hold exam sessions.

Can I use any reference books in the test?
Yes, you can use a copy of the IEE Wiring Regulations Seventeenth Edition.

How many different parts of the test are there?
There are 60 questions, which cover eight different sections of the syllabus.

Do I have a time limit for taking the test?
You have two hours to complete the test.

Do I need to be good at IT to do the test online?
No, the system is really easy to use, and you can practise before doing the test. There is also a practice GOLA test available to try on the IET website.

What happens if the computer crashes in the middle of my test?
This is unlikely, because of the way the system has been designed. If there is some kind of power or system failure, then your answers will be saved and you can continue on another machine if necessary.

Can people hack into the system and cheat?
There are lots of levels of security built into the system to ensure its safety. Also, each person gets a different set of questions, which makes it very difficult to cheat.

Can I change my answer?
Yes, you can change your answers quickly, easily and clearly at any time in the test up to the point where you end the exam. With any answers you feel less confident about, you can click the 'Flag' button, which means you can review these questions before you end the test.

How do I know how long I've got left to complete the test?
You can check the time remaining at any point during the exam by clicking on the 'Clock' icon in the tool bar. The time remaining will come up on the bottom right corner of the screen.

Notes

Is there only one correct A, B, C or D answer to multiple-choice questions?

Yes.

What happens if I don't answer all of the questions?

You should attempt to answer all of the questions. If you find a question difficult, mark it using the 'Flag' button and return to it later.

What grades of pass are there?

A Pass or a Fail.

When can I resit the test if I fail?

You can resit the exam at any time, and as soon as you and your tutor decide it is right for you, subject to the availability of the online examination.

Exam content

To help you to fully understand the exam content, this chapter is divided into the eight sections of the exam, which are in turn mapped to the seven parts and the appendices of BS 7671 (the *IEE Wiring Regulations*). The requirements of the examination are listed and references made to the relevant parts of BS 7671. One of the most useful parts of BS 7671 is its very comprehensive index. If you don't know where to look, use the index.

Section 1

Part 1 – Scope, object and fundamental principles

There are four questions on Part 1, which represent 7 per cent of the mark. Part 1 of BS 7671 contains Chapters 11, 12 and 13 and is the essence of the standard. It includes the scope, an important section as any installation for which requirements are being sought must be confirmed as being within the scope of BS 7671. It also includes the fundamental principles (Chapter 13). The rest of BS 7671, that is Parts 2 to 7, supports Part 1; Part 2 provides definitions of the terminology used throughout the standard and Parts 3 to 7 provide technical requirements intended to ensure that the electrical installation conforms with the fundamental principles of Chapter 13 of Part 1.

It is to be noted that departures from Parts 3 to 7 are allowed (Regulations 120.3 and 120.4), providing there is still compliance with Part 1. This allows the use of new materials and inventions not anticipated when the Regulations were published. However, departures from Parts 3 to 7, including the use of new materials and inventions, must not result in a degree of safety any less than that required by Part 1. Departures from Parts 3 to 7 must be recorded on the installation certificate.

To show that you are conversant with BS 7671, you are required to be able to:

1.1   Identify examples of installations in the scope of BS 7671 and particular requirements for specific installations and locations (for Scope, see Chapter 11 and Regulation 110.1)

1.2   Identify those installations that are excluded from BS 7671 (see Chapter 11 and Regulation 110.2)

1.3   Identify those statutory regulations that may be supported by BS 7671 (see Section 114 and Appendix 2)

1.4   State the requirements for installations in premises licensed under the statutory control of an authoritative body (see Section 115 and Appendix 2)

1.5 State the requirements regarding the fundamental principles relating to protection for safety, design, selection, erection, verification and certification of electrical installations (for fundamental principles, see Chapter 13).

Section 2

Part 2 – Definitions

There are three questions on Part 2, which represent 5 per cent of the mark. There are approximately 12 pages dedicated to the definitions used by the Regulations. Definitions that are particularly important, as their meanings in BS 7671 are often not the same as those in common English usage, include:

> **Exposed-conductive-part.** A conductive part of equipment which can be touched and which is not normally live, but which can become live when basic insulation fails.
>
> **Extraneous-conductive-part.** A conductive part liable to introduce a potential, generally Earth potential, and not forming part of the electrical installation.
>
> **Electrical equipment** (abbr: *Equipment*). Any item for such purposes as generation, conversion, transmission, distribution or utilisation of electrical energy, such as machines, transformers, apparatus, measuring instruments, protective devices, wiring systems, accessories, appliances and luminaires. (Be sure to note that 'equipment' is the abbreviation for 'electrical equipment'.)

Part 2 also includes useful technical information, such as the definitions of types of systems, eg '**TN-S system.** A system having separate neutral and protective conductors throughout the system.' A black and white drawing is provided as Fig 2.3 in Part 2 of the Regulations; for a colour drawing see opposite.

To show that you are conversant with BS 7671, you are required to be able to:
2.1 Use Part 2 of BS 7671.

Section 3

Part 3 – Assessment of general characteristics

There are five questions on Part 3, which represent 8 per cent of the mark. Part 3 requires a general assessment of the characteristics of the proposed installation. It also requires that the nature or characteristics of the supply be determined, as this plays an important part in the design of the installation. Consideration needs to be given to safety services and standby supplies. This part also includes requirements for the installation circuit arrangements.

TN-S system

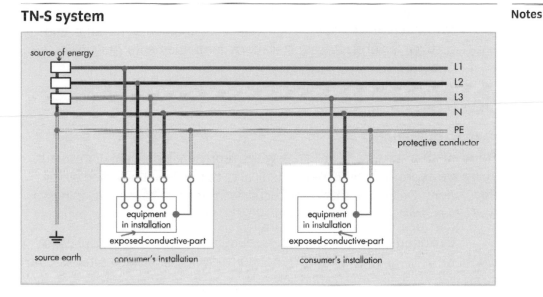

Separate neutral and protective conductors throughout the system.

The protective conductor (PE) is the metallic covering of the cable supplying the installations or a separate conductor.

All exposed-conductive-parts of an installation are connected to this protective conductor via the main earthing terminal of the installation.

To show that you are conversant with BS 7671, you are required to be able to:

3.1 State the need to determine maximum demand for an installation (see Regulation 311.1)

3.2 Determine the characteristics of a supply (see Regulation 313.1)

3.3 State the source (eg standby, external) and characteristics necessary for a supply (see Regulation 313.1)

3.4 State the need to divide an installation into suitable circuit arrangements (see Chapter 31, Section 314)

3.5 Determine the number and types of live conductors for installation circuits (see Regulation 312.2 and Appendix 6, Electrical Installation Certificate)

3.6 Identify those external influences that determine the selection of equipment and installation methods/techniques (see Regulation 301.1, Chapters 32 and 51 and Appendix 5)

3.7 State the need to consider compatibility and maintainability in the selection of equipment (see Regulation 301.1 and Chapters 33 and 34)

3.8 State the need for an assessment of each circuit regarding continuity of service (see Regulation 301.1 and Chapters 36 and 56).

Section 4

Part 4 – Protection for safety

There are 14 questions on Part 4, which represent 23 per cent of the mark. Part 4 contains some of the key chapters of BS 7671: Chapter 41, Chapter 42, Chapter 43 and Chapter 44.

To show that you are conversant with BS 7671, you are required to be able to:

4.1 Identify the differences between basic and fault protection (see Part 2 Definitions)

4.2 State means of protection against electrical shock by
    a   basic protection (see Sections 416 and 417)
    b   fault protection (see Sections 413, 415 and 418)
    c   both basic and fault protection (excluding IT) (see Sections 412 and 414)
    d   additional protection (see Section 415)

4.3 Describe how the requirements for shock protection are affected by
    a   the value of the external loop impedance (Z_e)
    b   compliance with $Z_s = Z_e + (R_1 + R_2)$
    c   compliance with Tables 41.1, 41.2, 41.3, 41.4, 41.5 and 41.6 (see Part 2 Definitions, Symbols used in the Regulations)

4.4 Identify the difference between overcurrent and fault current (see Part 2 Definitions and Chapter 43)

4.5 Describe means of protection against fire, burns and harmful thermal effects and identify precautions where particular risks of danger of fire exist (see Chapter 42, Section 422)

4.6 Identify the differences between overload currents, earth fault currents, short-circuit currents and shock currents (see Sections 433 and 434 and Part 2 Definitions)

4.7 Describe methods of overcurrent protection and the need for co-ordination with conductors and equipment (see Sections 432, 433 and 434)

4.8 State the requirements for protection against
    a   voltage disturbances
        (i)   overvoltage (see Chapter 44, Section 443)
        (ii)   undervoltage (see Chapter 44, Section 445)
    b   electromagnetic disturbances.

Section 5

Part 5 – Selection and erection of equipment

There are 15 questions on Part 5, which represent 25 per cent of the mark.

Part 5 includes requirements for voltage drop, warning notices and identification of conductors, as well as requirements for isolation and switching, earthing and bonding arrangements, protective conductors, lighting installations and generating sets.

To show that you are conversant with BS 7671, you are required to be able to:

5.1 Identify the need for compliance with British Standards, or harmonised European Standards and Codes of Practice (see Section 511)

5.2 State the effects that operational conditions and external influences have on the choice of installation methods, materials and equipment (see Section 512)

5.3 State the requirements for accessibility to equipment and connections (see Section 513)

5.4 Describe the need for, and the siting and wording requirements of, circuit charts, warning notices and labels (see Regulations 514.10 to 514.16)

5.5 State identification requirements for live and protective conductors (see Regulations 514.3, 514.4 and 514.5)

5.6 Describe the need to protect against mutual detrimental influences (see Section 515)

5.7 State the effect of each of the following on the selection and erection of wiring systems (see Chapter 52)
   a   types of wiring system (see Section 521)
   b   external influences (see Section 522)
   c   electrical connections (see Section 526)
   d   minimising the spread of fire (see Section 527)
   e   proximity to other services (see Section 528)
   f   maintainability (see Section 529)

5.8 Use the following to determine conductor size, and select appropriate cable from Appendix 4:
   a   design current
   b   overcurrent protection
   c   cable route
   d   correction factors
   e   voltage drop limitations
   f   diversity

5.9 State the difference between, and the requirements for
   a   isolation (see Part 2 Definitions and Regulation 537.2)
   b   switching off for mechanical maintenance (see Part 2 Definitions and Regulation 537.3)
   c   emergency switching (see Part 2 Definitions and Regulation 537.4)
   d   functional switching (see Part 2 Definitions and Regulation 537.5)

Notes

5.10   State the uses of monitoring
 a   insulation (see Regulations 538.1, 538.2 and 538.3)
 b   residual current (see Regulation 538.4)

5.11   Identify types of devices offering protection against risk of fire
(see Regulation 532)

5.12   Identify types of devices offering protection against overcurrent
(see Section 533)

5.13   Describe earthing arrangements to facilitate automatic disconnection
of supply (see Regulation 542.1)

5.14   State the recognised types of earth electrode and their applications
(see Regulation 542.2)

5.15   Determine, and select types and sizes of
 a   main equipotential bonding conductors (see Regulations 543.2
and 544.1)
 b   supplementary bonding conductors (see Regulations 543.2
and 544.2)
 c   circuit protective conductors (see Part 2 Definitions and
Section 543)
 d   earthing conductors (see Part 2 Definitions and Regulation 542.3)

5.16   Identify the requirements relevant to the installation of equipment
having High Protective Conductor Currents (see Regulation 543.7)

5.17   Describe the dangers in the use of equipment outside the
equipotential zone

5.18   State how electric shock protection is provided by
 a   RCD protected socket-outlets/circuits (see Regulation 411.3.3)
 b   the installation of an RCD in a TT system (see Regulations 411.5.2
and 531.4)

5.19   Identify the requirements for low voltage generating sets
(see Section 551)

5.20   Identify the requirements for rotating machines (see Section 552)

5.21   State the requirements for accessories such as plug and socket-outlets
and cable couplers (see Regulations 533.1 and 533.2)

5.22   Identify the requirements for current using equipment such as
electrode boilers and heating cables (see Section 554)

5.23   State the requirements for transformers (see Section 555)

5.24   Identify the requirements for the selection and erection of luminaires
and lighting installations, including highway power supplies and street
furniture (see Chapter 55 and Section 559)

5.25   State the requirements for supplies for safety services and their
associated circuits and cables (see Chapter 56).

Section 6

Part 6 – Inspection and testing

There are six questions on Part 6, which represent 10 per cent of the mark.

To show that you are conversant with BS 7671, you are required to be able to:

6.1 Identify the general requirements for inspection and testing of installations (see Section 610)

6.2 State the need for diagrams, charts or tables to be available prior to the verification process (see Regulation 610.2)

6.3 Identify the items to be checked during the inspection process (see Regulation 611.3)

6.4 State the tests which may need to be carried out on initial verification and period inspection (see Regulations 612.2 to 612.14)

6.5 Compare test results with relevant criteria (see Regulation 612.1)

6.6 State the requirements for the issue of
  a   Electrical Installation Certificate (see Sections 631 and 632)
  b   Minor Works Certificate (see Sections 631 and 633)
  c   Periodic Inspection Report (see Sections 631 and 634).

Section 7

Part 7 – Special installations or locations

There are ten questions on Part 7, which represent 17 per cent of the mark. Part 7 includes additional requirements or altered requirements to those in Parts 3 to 5. The requirements of the rest of BS 7671 apply to the special locations, unless they are amended or altered by the particular requirements in Part 7.

To show that you are conversant with BS 7671, you are required to be able to:

7.1 State the requirements for safety measures in locations containing a bath or shower (see Section 701)

7.2 State the special precautions that must be applied regarding swimming pools and other basins (see Section 702)

7.3 State the requirements that must be applied to rooms or cabins containing sauna heaters (see Section 703)

7.4 Identify the requirements relevant to temporary installations within construction and demolition sites (see Section 704)

7.5 Identify the requirements relevant to installations within agricultural and horticultural premises (see Section 705)

7.6 Identify the requirements for electrical installations in caravans, motor caravans and caravan parks (see Sections 708 and 721)

7.7 Identify the requirements for conducting locations with restricted movement (see Section 706)

Notes

7.8 Identify the requirements for marinas and similar locations (see Section 709)

7.9 Identify the requirements for exhibitions, shows and stands (see Section 711)

7.10 Identify the requirements for locations containing solar photovoltaic power supply systems (see Section 712)

7.11 Identify the requirements for mobile or transportable units (see Section 717)

7.12 Identify the requirements for temporary electrical installations for structures, amusement devices and booths at fairgrounds, amusement parks and circuses (see Section 740)

7.13 Identify the requirements for locations containing floor and ceiling heating systems (see Section 753).

Section 8

There are three questions on the Appendices of BS 7671, which represent 5 per cent of the mark.

To show that you are conversant with BS 7671, you are required to be able to:

8.1 Apply relevant information/data within Appendices:

a British Standards to which reference is made in the Regulations (see Appendix 1)

b Statutory regulations and associated memoranda (see Appendix 2)

c Time/current characteristics of overcurrent protective devices and residual current devices (see Appendix 3)

d Current-carrying capacity and voltage drop for cables and flexible cords (see Appendix 4)

e Classification of external influences (see Appendix 5)

f Model forms for certification and reporting (see Appendix 6)

g Harmonised cable core colours (see Appendix 7)

h Busbar trunking and powertrack systems (see Appendix 8)

i Multiple source, d.c. and other systems (see Appendix 9)

j Protection of conductors in parallel (see Appendix 10)

k Harmonic currents (see Appendix 11)

l Voltage drop (see Appendix 12)

m Measurement of impedance of floors and walls (see Appendix 13)

n Measurement of earth loop impedance (see Appendix 14)

o Ring and radial final circuit arrangements (see Appendix 15).

Tips from the examiner

The following tips are intended to aid confident test performance. Some are more general and would apply to most exams. Others are more specific, either because of the format of this test (multiple choice) or the nature of the subject.

✔ If you rarely use a computer, try to get some practice beforehand. You need to be able to use a mouse to move a cursor arrow around a computer screen, as you will use the cursor to click on the correct answer in the exam.

✔ Take a copy of *BS 7671: 2008 Requirements for Electrical Installations, IEE Wiring Regulations Seventeenth Edition* into the exam. Take the time to familiarise yourself with the structure and content of this publication.

✔ Make the most of the course you will attend before taking the exam. Try to attend all sessions and be prepared to devote time outside the class to revise for the exam.

✔ On the day of the exam, allow plenty of time for travel to the centre and arrive at the place of the exam at least ten minutes before it's due to start so that you have time to relax and get into the right frame of mind.

✔ Listen carefully to the instructions given by the invigilator.

✔ Read the question and every answer before making your selection. Do not rush – there should be plenty of time to answer all the questions.

✔ Look at the exhibits where instructed. Remember, an exhibit supplies you with information that is required to answer the question.

✔ Attempt to answer all the questions. If a question is not answered, it is marked as wrong. Making an educated guess improves your chances of choosing the correct answer. Remember, if you don't select an answer, you will definitely get no marks for that question.

✔ The order of the exam questions follows the order of the *IEE Wiring Regulations Seventeenth Edition* publication. Therefore, look for the answers to early questions at the front of the book and progress through it as you work through the exam questions.

Notes

✔ Don't worry about answering the questions in the order in which they appear in the exam. Choose the 'Flag' option on the tool bar to annotate the questions you want to come back to. If you spend too much time on questions early on, you may not have time to answer the later questions, even though you know the answers.

✔ Although not absolutely necessary, some candidates find it useful to bring a basic, non-programmable calculator with square-root ($\sqrt{\ }$) and square (x^2) functions to the exam.

✔ If you are having trouble finding the regulation for a particular question, look for the subject in the index of the *IEE Wiring Regulations Seventeenth Edition*. Using the index should minimise the time it takes you to find the relevant topic.

✔ It is not recommended that you memorise any of the material presented here in the hope it will come up in the exam. The exam questions featured in this book will help you to gauge the kinds of questions that might be asked. It is highly unlikely you will be asked any identical questions in the exam, but you may see variations on certain themes.

Exam practice 1

Exam practice 1

Sample test 1

The sample test below has 60 questions, the same number as the online exam, and its structure follows that of the online exam. The test appears first without answers, so you can use it as a mock exam. It is then repeated with worked-through answers and extracts from the *IEE Wiring Regulations*. Finally, there is an answer key for easy reference.

Answer the questions by filling in the circle next to your chosen option.

Section 1

1 BS 7671 applies to

- a   equipment on board ships
- b   lightning protection of buildings
- c   lift installations
- d   prefabricated buildings.

2 BS 7671 provides requirements for safety against the risk of

- a   electric shock on an aircraft
- b   shock currents on board ships
- c   fire on offshore installations
- d   shock currents in electrical installations.

3 Which one of the following is <u>not</u> a statutory regulation?

- a   Electricity at Work Regulations 1989 as amended
- b   The Supply of Machinery (Safety) Regulations 1992 as amended
- c   Requirements for Electrical Installations (BS 7671)
- d   Agricultural (Stationary Machinery) Regulations

4 BS 7671 identifies that the cross-sectional area of a conductor shall be determined by

- a   the admissible maximum temperature
- b   the nominal voltage
- c   voltage tolerances
- d   the earthing system.

Section 2

Notes

5   A corridor containing supporting structures for cables and joints and/or other elements of wiring systems, the dimensions of which allow persons to pass freely throughout the entire length, is known as

○ a   an access pathway
○ b   a cable tunnel
○ c   an access throughway
○ d   cable ducting.

6   The algebraic sum of the currents in the live conductors of a circuit at a point in the electrical installation is known as the

○ a   residual current
○ b   harmonic current
○ c   line current
○ d   neutral current.

7   An assembly of PV arrays is defined as a

○ a   PV cell
○ b   PV array cable
○ c   PV generator
○ d   PV a.c. module.

Section 3

8   With reference to the nature of the supply, which one of the following can be determined by calculation, enquiry or measurement?

○ a   The maximum demand of the installation
○ b   The rating of the circuit protective device
○ c   The prospective short-circuit current at the origin of the installation
○ d   The csa of the tails

Notes

9 Every installation should be divided into individual circuits to

○ a   prevent faults developing
○ b   provide protection against electric shock
○ c   minimize inconvenience in the event of a fault
○ d   ease installation.

10 Diversity may be taken into account when considering

○ a   maximum demand of the installation
○ b   a TN-C-S system
○ c   the prospective short-circuit fault current
○ d   the number of final circuits.

11 Which one of the following is <u>not</u> a characteristic of the supply?

○ a   The nature of the current and frequency
○ b   The earth fault loop impedance external to the installation
○ c   Main switch current rating
○ d   The nominal voltage

12 BS 7671 requires designers to take into account the frequency and quality of maintenance an installation can reasonably be expected to receive when

○ a   assessing staff numbers
○ b   inspecting and testing
○ c   selecting staff
○ d   specifying or selecting equipment reliability.

Section 4

13 Which one of the following is <u>not</u> part of the requirements for fault protection?

○ a   Protective earthing
○ b   Protective equipotential bonding
○ c   Automatic disconnection
○ d   Protection by insulation of live parts

14 The maximum disconnection time for a TN system with a nominal voltage of 400 V a.c. to Earth is

- a 0.2 second
- b 0.4 second
- c 0.5 second
- d 5 seconds.

15 For a TT system, which one of the following conditions should be fulfilled for each circuit protected by an RCD?

- a $R_A \times I_{\Delta n} \leq 50\,V$
- b $R_A \times I_{\Delta n} \geq 50\,V$
- c $R_A \times I_d \leq 50\,V$
- d $R_A \times I_d \geq 50\,V$

16 A 32 A type B circuit-breaker is used to give a disconnection time of 5 seconds in a reduced low voltage system with a nominal voltage to Earth (U_O) of 55 V. What is the maximum value of earth fault loop impedance (Z_S)?

- a $0.44\,\Omega$
- b $0.34\,\Omega$
- c $0.17\,\Omega$
- d $0.09\,\Omega$

17 A SELV source can be derived from which one of the following?

- a Double-wound transformer
- b Autotransformer
- c Safety isolating transformer
- d Step-up transformer

18 Which one of the following cannot be used as basic protection?

- a Insulation of live parts
- b Barriers or enclosures
- c Protective earthing and bonding
- d Obstacles

Notes

19 In order to provide basic protection, a horizontal top surface of a barrier or enclosure that is readily accessible shall provide a minimum degree of protection of

○ a   IPXXA or IP1X
○ b   IPXXB or IP2X
○ c   IPXXC or IP3X
○ d   IPXXD or IP4X.

20 Where arcs, sparks or particles at high temperature may be emitted by fixed equipment in normal service, the equipment shall be

○ a   totally enclosed in arc-resistant material
○ b   protected by a 30 mA RCD
○ c   enclosed to at least IP55
○ d   accessible only by use of a key or tool.

21 Except where otherwise recommended by the manufacturer, spotlights and projectors rated at over 100 W and up to 300 W shall be installed at a minimum distance from combustible materials of

○ a   0.5 m
○ b   0.6 m
○ c   0.8 m
○ d   1.0 m.

22 In locations with increased risks of fire, motors which are automatically or remotely controlled, or which are not continuously supervised, shall be protected against excessive temperature by

○ a   a protective device that is automatically reset
○ b   a protective device with manual reset
○ c   electronic monitoring equipment that resets
○ d   electronic monitoring equipment that restarts the motor.

23 Where particular risks of fire exist, the classification for high density occupation areas with easy conditions of evacuation is

○ a   BD1
○ b   BD2
○ c   BD3
○ d   BD4.

24 When considering protection against overload, the symbol for the current ensuring effective operation of the protective device in the conventional time is

○ a I_b
○ b I_z
○ c I_n
○ d I_2.

25 For protection against overvoltage a 230 V electricity meter should have an impulse withstand of

○ a   6 kV
○ b   4 kV
○ c   2.5 kV
○ d   1.5 kV.

26 What is the impulse category of equipment that is part of the fixed electrical installation and other equipment where a high degree of availability is expected?

○ a   I
○ b   II
○ c   III
○ d   IV

Section 5

27 If the construction of equipment is unsuited to the external influences of its location, it should be

○ a   given a plastic coating
○ b   given a zinc finish
○ c   provided with additional protection during erection
○ d   supplied by SELV only.

28 A permanent label to BS 951 bearing the words 'Safety Electrical Connection – Do Not Remove' is <u>not</u> required at

○ a   the connection of every earthing conductor to an earth electrode

○ b   the point of connection of every bonding conductor to an extraneous-conductive-part

○ c   the main earth terminal, where separate from the main switchgear

○ d   a main earthing bar contained within switchgear.

29 A functional switch has to be provided for each part of the circuit

○ a   that may require independent control

○ b   1200 mm from the floor

○ c   for safe isolation

○ d   for emergency switching purposes.

30 A firefighter's switch shall be provided in the low voltage circuit supplying exterior electrical installations and interior discharge lighting operating at

○ a   a voltage exceeding low voltage

○ b   low voltage

○ c   voltage band II

○ d   medium voltage.

31 A plug and socket-outlet may be used for switching off for mechanical maintenance as long as it does not have a rating exceeding

○ a   13 A

○ b   16 A

○ c   32 A

○ d   45 A.

32 If an area within an installation undergoes a 10 °C rise in ambient temperature, the effect on the current-carrying capacity of cables will be to

○ a   decrease the value of I_z

○ b   increase the value of I_z

○ c   leave I_z unchanged

○ d   increase the fault current by 10 per cent.

33 A multicore 70 °C thermoplastic cable with 2.5 mm² conductors supplies a single-phase load of 20 A at 230 V a.c. over a distance of 22 metres. The voltage drop in the cable will be

- ○ a   6 V
- ○ b   6.6 V
- ○ c   7.92 V
- ○ d   9.2 V.

34 Where a 3-core cable, with cores coloured brown, black and grey, is used as a switch wire for two-way or intermediate control, the terminations of the conductors shall be identified using

- ○ a   red, blue and yellow tape
- ○ b   black tape only on each core
- ○ c   brown tape on the black and grey cores
- ○ d   self-colour tape only.

35 Which one of the following cannot be used as an earth electrode?

- ○ a   Earth plates
- ○ b   Welded reinforcement of concrete embedded in the earth
- ○ c   Earth tapes
- ○ d   Gas and water utility pipes

36 An earthing conductor buried in the ground is protected against corrosion by a sheath, but is not protected against mechanical damage. The minimum size copper conductor that may be installed is

- ○ a   2.5 mm²
- ○ b   16 mm²
- ○ c   25 mm²
- ○ d   50 mm².

Notes

37 Assuming that both the line and protective conductors are of the same material, for a line conductor of 10 mm^2 if the protective conductor is to be selected, the minimum tabulated cross-sectional area of its associated protective conductor is

○ a   6 mm^2

○ b   10 mm^2

○ c   16 mm^2

○ d   35 mm^2.

38 A radial final circuit feeding socket-outlets supplying several items of data processing equipment has a total protective conductor current in normal service of 18 mA. This circuit must have a high integrity protective conductor

○ a   of cross-sectional area less than 1 mm^2

○ b   connected as a ring

○ c   controlled by an isolator

○ d   enclosed in insulated conduit only.

39 A static type uninterruptible power supply source shall comply with

○ a   BS 3036

○ b   BS 1361

○ c   BS EN 60898

○ d   BS EN 62040.

40 Where an autotransformer is connected to a circuit having a neutral conductor, the common terminal of the winding shall be connected to the

○ a   neutral conductor

○ b   line conductor

○ c   protective conductor

○ d   bonding conductor.

41 **When selecting wiring systems for safety services, the type of cable that should be used in fire conditions should comply with**

- a   BS 5467
- b   BS 6231
- c   BS 7211
- d   BS EN 50362.

Section 6

42 **When carrying out a visual inspection of an electrical installation, which one of the following does <u>not</u> have to be verified?**

- a   The methods of protection against electric shock
- b   The electricity supplier
- c   The connection of conductors
- d   The presence of undervoltage protective devices

43 **Certain information must be made available to persons carrying out inspection and testing of an installation before the testing commences. One such item of information would be**

- a   the name of the client
- b   the name of the person who designed the installation
- c   the length of cable runs in the installation
- d   any circuit or equipment vulnerable to a typical test.

44 **When carrying out an inspection of a new installation, it is <u>not</u> necessary to verify the**

- a   total earth fault loop impedance for each circuit
- b   connection of conductors
- c   methods of protection against electric shock
- d   presence of diagrams, instructions and similar information.

45 **An insulation resistance test is to be carried out on a 3-phase 400 V circuit. The test voltage and minimum acceptable reading would be**

- a   250 V a.c. and 0.5 MΩ
- b   500 V d.c. and 0.5 MΩ
- c   500 V d.c. and 1 MΩ
- d   800 V d.c. and 0.5 Ω.

Notes

46 When an addition is made to an existing installation, the contractor shall record on the Electrical Installation Certificate or the Minor Electrical Installation Works Certificate any

- ○ a changes in ownership
- ○ b records of repair over the last five years
- ○ c defects in the existing installation
- ○ d voltage drop in the longest circuit.

47 After completion of a periodic inspection, the completed documentation shall be given to the

- ○ a person ordering the inspection
- ○ b local authority
- ○ c insurance company
- ○ d main contractor.

Section 7

48 In a room containing a bath, electrical equipment installed in zone 0 shall have a degree of protection of at least

- ○ a IPX5
- ○ b IP5X
- ○ c IP7X
- ○ d IPX7.

49 A flush downlighter in the ceiling less than 2.5 m height above floor level over a bath will need to fulfil the requirements of

- ○ a zone 2
- ○ b zone 1
- ○ c zone 0
- ○ d all zones.

50 Which of the following protective measures is permitted in a room containing a bath or shower?

- ○ a Obstacles
- ○ b Placing out of reach
- ○ c Automatic disconnection of supply
- ○ d Non-conducting location

51 In a swimming pool or other basin, the metallic covering or sheath of a wiring system in zones 0, 1 or 2 shall

- ○ a   not be used
- ○ b   not be earthed
- ○ c   be connected to the supplementary bonding
- ○ d   have reinforced insulation.

52 The requirements of Section 704 of BS 7671 apply to

- ○ a   cloakrooms
- ○ b   offices
- ○ c   construction and demolition site installations
- ○ d   toilets

53 In locations where livestock is kept, for all circuits other than socket-outlet circuits, an RCD shall be provided with a rating not exceeding

- ○ a   500 mA
- ○ b   300 mA
- ○ c   100 mA
- ○ d   30 mA.

54 Equipment on a pontoon in a marina that is subject to impact to level AG2 should have a mechanical protection code of

- ○ a   IPX4
- ○ b   IPX8
- ○ c   IP55
- ○ d   IK08.

55 For marinas, the classification of external influence which does <u>not</u> need to be considered is

- ○ a   AD
- ○ b   AE
- ○ c   AF
- ○ d   AP.

56 In caravans, each final circuit shall be protected against overcurrent by a device that disconnects

- ○ a all live conductors in that circuit
- ○ b the line conductors only
- ○ c line, neutral and protective conductors
- ○ d the caravan and site supply system.

57 Electric dodgems shall only be operated at voltages not exceeding

- ○ a 120 V a.c. or 50 V d.c.
- ○ b 50 V a.c. or 120 V d.c.
- ○ c 110 V a.c. reduced low voltage
- ○ d 1000 V a.c. or 1500 V d.c.

Section 8

58 The requirements of the licensing authority should be ascertained for

- ○ a installations on caravan parks
- ○ b temporary supplies
- ○ c types of earthing system to be used
- ○ d the design of the installation.

59 I_2 can be greater than 1.45 I_z when the overcurrent device

- ○ a is providing overload protection
- ○ b is a circuit-breaker
- ○ c is rated at over 100 A
- ○ d is providing fault current protection only.

60 External influences coded BE are classified as

- ○ a nature of processed or stored materials
- ○ b conditions of evacuation in an emergency
- ○ c movement of air
- ○ d capability of persons.

Questions and answers

Notes

The questions in sample test 1 are repeated below with worked-through answers and extracts from the *IEE Wiring Regulations* where appropriate.

Section 1

1   BS 7671 applies to

- ○ a   equipment on board ships
- ○ b   lightning protection of buildings
- ○ c   lift installations
- ◉ d   prefabricated buildings.

Answer d

See Part 1, Chapter 11, Scope. Option d, prefabricated buildings, is listed in Regulation 110.1, item (vi) as being within the scope. Options a, b and c – equipment on board ships, lightning protection of buildings and lift installations – are listed in Regulation 110.2 as being excluded from the scope of the Regulations.

2   BS 7671 provides requirements for safety against the risk of

- ○ a   electric shock on an aircraft
- ○ b   shock currents on board ships
- ○ c   fire on offshore installations
- ◉ d   shock currents in electrical installations.

Answer d

Regulation 110.2, Exclusions from scope, advises that options a, b and c are outside the scope.

Regulation 131.1 (i) shock currents in Chapter 13, Fundamental Principles, confirms that the requirements of BS 7671 are intended to provide for the safety of persons and livestock against the risk of injury from shock currents.

3   Which one of the following is <u>not</u> a statutory regulation?

- ○ a   Electricity at Work Regulations 1989 as amended
- ○ b   The Supply of Machinery (Safety) Regulations 1992 as amended
- ◉ c   Requirements for Electrical Installations (BS 7671)
- ○ d   Agricultural (Stationary Machinery) Regulations

Notes

Answer c

Despite having the subtitle *IEE Wiring Regulations*, the *Requirements for Electrical Installations (BS 7671)* is not a statutory regulation: see Regulation 114.1. BS 7671 is a British Standard and encompasses the recommendations of the IEE.

4   BS 7671 identifies that the cross-sectional area of a conductor shall be determined by

- ⦿ a   the admissible maximum temperature
- ○ b   the nominal voltage
- ○ c   voltage tolerances
- ○ d   the earthing system.

Answer a

Design is considered in Section 132, and Regulation 132.6 (Cross-sectional area of conductors) is particularly relevant. Option a – 'the admissible maximum temperature' – is listed as being a factor in determining the cross-sectional area. Voltage (options b and c) may affect the type of wiring (see Regulation 132.7). Option d may affect the method of protection against electric shock (see Chapter 41).

132.6 Cross-sectional area of conductors

The cross-sectional area of conductors shall be determined for both normal operating conditions and, where appropriate, for fault conditions according to:

(i)   the admissible maximum temperature

(ii)   the voltage drop limit

(iii)   the electromechanical stresses likely to occur due to short-circuit and earth fault currents

(iv)   other mechanical stresses to which the conductors are likely to be exposed

(v)   the maximum impedance for operation of short-circuit and earth fault protection

(vi)   the method of installation

(vii) harmonics

(viii) thermal insulation.

Section 2

5   **A corridor containing supporting structures for cables and joints and/or other elements of wiring systems, the dimensions of which allow persons to pass freely throughout the entire length, is known as**

○ a   an access pathway
◉ b   a cable tunnel
○ c   an access throughway
○ d   cable ducting.

Answer b
Options a and c are not defined in BS 7671. Options b and d are both defined in Part 2 Definitions, which show that option b is correct.

Cable ducting. An enclosure of metal or insulating material, other than conduit or cable trunking, intended for the protection of cables which are drawn in after erection of the ducting.

Cable tunnel. A corridor containing supporting structures for cables and joints and/or other elements of wiring systems and whose dimensions allow persons to pass freely throughout the entire length.

6   **The algebraic sum of the currents in the live conductors of a circuit at a point in the electrical installation is known as the**

◉ a   residual current
○ b   harmonic current
○ c   line current
○ d   neutral current.

Answer a
The answer is found in Part 2 Definitions, where option a is defined. Option b, harmonic currents, have frequencies that are multiples of the fundamental. Options c and d, line and neutral currents, are the currents flowing in line and neutral conductors. They are not defined.

Residual current. Algebraic sum of the currents in the live conductors of a circuit at a point in the electrical installation.

Notes

7 An assembly of PV arrays is defined as a

- ○ a   PV cell
- ○ b   PV array cable
- ● c   PV generator
- ○ d   PV a.c. module.

Answer c

The answer is found in Part 2 Definitions, where option c is clearly defined as an 'assembly of PV arrays'.

Section 3

8 With reference to the nature of the supply, which one of the following can be determined by calculation, enquiry or measurement?

- ○ a   The maximum demand of the installation
- ○ b   The rating of the circuit protective device
- ● c   The prospective short-circuit current at the origin of the installation
- ○ d   The csa of the tails

Answer c

Regulation 313.1 allows that certain of the supply characteristics may be determined either by enquiry of the electricity distributor/supplier or by measurement on-site or, in some circumstances such as a proposed distribution system, by calculation. The prospective short-circuit current at the origin is normally determined by enquiry of the distributor, but could also be measured on-site in some circumstances. Options a, b and d are not listed as characteristics of the supply.

9 Every installation should be divided into individual circuits to

- ○ a   prevent faults developing
- ○ b   provide protection against electric shock
- ● c   minimize inconvenience in the event of a fault
- ○ d   ease installation.

Answer c

Regulation 314.1, item (i), shows that the answer is option c. Dividing an installation into individual circuits does not prevent faults developing (option a) or provide protection against electric shock (option b). It may ease installation (option d), but that is not the reason that BS 7671 requires it.

314.1 Every installation shall be divided into circuits, as necessary, to:

(i) **avoid hazards and minimize inconvenience in the event of a fault**

(ii) facilitate safe inspection, testing and maintenance (see also Section 537)

(iii) take account of danger that may arise from the failure of a single circuit such as a lighting circuit

(iv) reduce the possibility of unwanted tripping of RCDs due to excessive protective conductor currents produced by equipment in normal operation

(v) mitigate the effects of electromagnetic interferences (EMI)

(vi) prevent the indirect energizing of a circuit intended to be isolated.

10 Diversity may be taken into account when considering

- ⦿ a maximum demand of the installation
- ○ b a TN-C-S system
- ○ c the prospective short-circuit fault current
- ○ d the number of final circuits.

Answer a

Diversity between the instant of peak demand of the various loads of an installation may mean that the maximum demand at any time is less than the simple sum of the loads (see Regulation 311.1). Diversity must be taken into account when determining maximum demands, otherwise switchgear and cables may be oversized.

311.1 For economic and reliable design, the maximum demand of an installation shall be assessed. In determining the maximum demand of an installation or part thereof, diversity may be taken into account.

11 Which one of the following is <u>not</u> a characteristic of the supply?

- ○ a The nature of the current and frequency
- ○ b The earth fault loop impedance external to the installation
- ⦿ c Main switch current rating
- ○ d The nominal voltage

Answer c

Regulation 313.1 shows that options a, b and d are characteristics of the supply, therefore option c must be the answer. Main switch current rating is a feature of the installation, rather than a characteristic of the supply.

313.1 The following characteristics of the supply or supplies, from whatever source, and the normal range of those characteristics where appropriate, shall be determined by calculation, measurement, enquiry or inspection:

(i) The nominal voltage(s) and its characteristics including harmonic distortion

(ii) The nature of the current and frequency

(iii) The prospective short-circuit current at the origin of the installation

(iv) The earth fault loop impedance of that part of the system external to the installation, Z_e

(v) The suitability for the requirements of the installation, including the maximum demand

(vi) The type and rating of the overcurrent protective device(s) acting at the origin of the installation.

12 BS 7671 requires designers to take into account the frequency and quality of maintenance an installation can reasonably be expected to receive when

○ a   assessing staff numbers
○ b   inspecting and testing
○ c   selecting staff
◉ d   specifying or selecting equipment reliability.

Answer d

Regulation 341.1 requires the designer and installer to take into account the likely maintenance of the installation when selecting equipment. Options a and c are not listed as factors for consideration. Inspection and testing **is** mentioned, but inspection and testing are not a designer's function.

341.1 An assessment shall be made of the frequency and quality of maintenance the installation can reasonably be expected to receive during its intended life. The person or body responsible for the operation and/or maintenance of the installation shall be consulted. Those characteristics are to be taken into account in applying the requirements of Parts 4 to 7 so that, having regard to the frequency and quality of maintenance expected:

(i) any periodic inspection and testing, maintenance and repairs likely to be necessary during the intended life can be readily and safely carried out, and

(ii) the effectiveness of the protective measures for safety during the intended life shall not diminish, and

(iii) the reliability of equipment for proper functioning of the installation is appropriate to the intended life.

Section 4

13 Which one of the following is <u>not</u> part of the requirements for fault protection?

○ a Protective earthing
○ b Protective equipotential bonding
○ c Automatic disconnection
◉ d Protection by insulation of live parts

Answer d
The methods of fault protection for the protective measure of automatic disconnection of supply are found in Regulation 411.3. Protection by basic insulation of live parts (option d) is not listed under 411.3 as it is a method of basic protection.

14 The maximum disconnection time for a TN system with a nominal voltage of 400 V a.c. to Earth is

◉ a 0.2 second
○ b 0.4 second
○ c 0.5 second
○ d 5 seconds.

Answer a
You can find the maximum disconnection times for TN systems listed in Table 41.1. Note that the 400 V in the question is a.c. voltage to Earth, U_0, and not phase to phase, hence the answer is 0.2 second.

Table 41.1 Maximum disconnection times

System	$50\,V < U_0 \leq 120\,V$ seconds		$120\,V < U_0 \leq 230\,V$ seconds		$230\,V < U_0 \leq 400\,V$ seconds		$U_0 > 400\,V$ seconds	
	a.c.	d.c.	a.c.	d.c.	a.c.	d.c.	a.c.	d.c.
TN	0.8	NOTE 1	0.4	5	**0.2**	0.4	0.1	0.1
TT	0.3	NOTE 1	0.2	0.4	0.07	0.2	0.04	0.1

From the *IEE Wiring Regulations*, Table 41.1, page 46

Notes

15 For a TT system, which one of the following conditions should be fulfilled for each circuit protected by an RCD?

- ⦿ a $R_A \times I_{\Delta n} \leq 50$ V
- ◯ b $R_A \times I_{\Delta n} \geq 50$ V
- ◯ c $R_A \times I_d \leq 50$ V
- ◯ d $R_A \times I_d \geq 50$ V

Answer a

You can find the requirements for protection by automatic disconnection of supply in TT systems stated in Regulation 411.5, and Regulation 411.5.3 provides the answer. This requirement ensures that, if the fault current to earth is insufficient to operate the protective device, the difference in potential between the exposed-conductive-part and true earth will not exceed 50 V. The equation below means that R_A multiplied by $I_{\Delta n}$ must be less than or equal to 50 V.

411.5.3 Where an RCD is used for earth fault protection, the following conditions shall be fulfilled:
(i)   The disconnection time shall be that required by Regulation 411.3.2.2 or 411.3.2.4, and
(ii) $R_A \times I_{\Delta n} \leq 50$ V

where:
R_A is the sum of the resistances of the earth electrode and the protective conductor connecting it to the exposed-conductive-parts (in ohms).
$I_{\Delta n}$ is the rated residual operating current of the RCD.

16 A 32 A type B circuit-breaker is used to give a disconnection time of 5 seconds in a reduced low voltage system with a nominal voltage to Earth (U_o) of 55 V. What is the maximum value of earth fault loop impedance (Z_s)?

- ◯ a $0.44 \, \Omega$
- ⦿ b $0.34 \, \Omega$
- ◯ c $0.17 \, \Omega$
- ◯ d $0.09 \, \Omega$

Answer b

Reference to the index for 'Reduced low voltage system' will direct you to Regulation 411.8 and Table 41.6. See column 2 of Table 41.6, which gives a maximum value for a 32 A type B circuit-breaker of $0.34 \, \Omega$.

Table 41.6   Maximum earth fault loop impedance (Z_s) for 5 s disconnection time and U_0 of 55 V (single-phase) and 63.5 V (three-phase)

Notes

	Circuit-breakers to BS EN 60898 and the overcurrent characteristics of RCBOs to BS EN 61009-1 Type						General purpose (gG) fuses to BS 88-2.2 and BS 88-6	
	B		C		D			
U_0 (Volts)	55	63.5	55	63.5	55	63.5	55	63.5
Rating (amperes)				Z_s ohms				
6	1.83	2.12	0.92	1.07	0.47	0.53	3.20	3.70
10	1.10	1.27	0.55	0.64	0.28	0.32	1.77	2.05
16	0.69	0.79	0.34	0.40	0.18	0.20	1.00	1.15
20	0.55	0.64	0.28	0.32	0.14	0.16	0.69	0.80
25	0.44	0.51	0.22	0.26	0.11	0.13	0.55	0.63
32	**0.34**	0.40	0.17	0.20	0.09	0.10	0.44	0.51
40	0.28	0.32	0.14	0.16	0.07	0.08	0.32	0.37
50	0.22	0.25	0.11	0.13	0.06	0.06	0.25	0.29
63	0.17	0.20	0.09	0.10	0.04	0.05	0.20	0.23
80	0.14	0.16	0.07	0.08	0.04	0.04	0.14	0.16
100	0.11	0.13	0.05	0.06	0.03	0.03	0.10	0.12
125	0.09	0.10	0.04	0.05	0.02	0.03	0.08	0.09

From the *IEE Wiring Regulations*, Table 41.6, page 54

Notes

17 A SELV source can be derived from which one of the following?

○ a   Double-wound transformer
○ b   Autotransformer
⊙ c   Safety isolating transformer
○ d   Step-up transformer

Answer c

There are three basic requirements for SELV: (i) the voltage does not exceed extra-low voltage, (ii) the supply is from a safety source and (iii) the live parts are insulated from Earth.

The sources for SELV are specified in Regulation 414.3. Only option c – 'safety isolating transformer' – is listed.

414.3   Sources for SELV and PELV

The following sources may be used for SELV and PELV systems:

(i)   **A safety isolating transformer in accordance with BS EN 61558-2-6**

(ii)   A source of current providing a degree of safety equivalent to that of the safety isolating transformer specified in (i) (eg motor-generator with windings providing equivalent isolation)

(iii)   An electrochemical source (eg a battery) or another source independent of a higher voltage circuit (eg a diesel-driven generator)

(iv)   Certain electronic devices complying with appropriate standards, where provisions have been taken to ensure that, even in the case of an internal fault, the voltage at the outgoing terminals cannot exceed the values specified in Regulation 414.1.1. Higher voltages at the outgoing terminals are, however, permitted if it is ensured that, in case of contact with a live part or in the event of a fault between a live part and an exposed-conductive-part, the voltage at the output terminals is immediately reduced to the value specified by Regulation 414.1.1 or less.

A mobile source supplied at low voltage, eg a safety isolating transformer or a motor-generator, shall be selected and erected in accordance with the requirements for protection by the use of double or reinforced insulation (see Section 412).

18 Which one of the following <u>cannot</u> be used as basic protection?

- ○ a   Insulation of live parts
- ○ b   Barriers or enclosures
- ◉ c   Protective earthing and bonding
- ○ d   Obstacles

Answer c

Options a, b and d – insulation of live parts, barriers or enclosures, and obstacles – all provide basic protection (see Sections 416 and 417). Option c, protective earthing and protective bonding, cannot be used as basic protection – these provide fault protection where the protective measure automatic disconnection of supply is employed.

19 In order to provide basic protection, a horizontal top surface of a barrier or enclosure that is readily accessible shall provide a minimum degree of protection of

- ○ a   IPXXA or IP1X
- ○ b   IPXXB or IP2X
- ○ c   IPXXC or IP3X
- ◉ d   IPXXD or IP4X.

Answer d

Although BS 7671 does not list all the IP codes, reference to some of them, as appropriate, is made. The most frequently quoted are IPXXB or IP2X. However, in this case the correct answer is IPXXD or IP4X (see Regulation 416.2.2).

416.2.2 A horizontal top surface of a barrier or enclosure which is readily accessible shall provide a degree of protection of at least IPXXD or IP4X.

20 Where arcs, sparks or particles at high temperature may be emitted by fixed equipment in normal service, the equipment shall be

- ◉ a   totally enclosed in arc-resistant material
- ○ b   protected by a 30 mA RCD
- ○ c   enclosed to at least IP55
- ○ d   accessible only by use of a key or tool.

Answer a

This comes under Section 421, Protection against fire caused by electrical equipment, and in particular, Regulation 421.3.

421.3 Where arcs, sparks or particles at high temperature may be emitted by fixed equipment in normal service, the equipment shall meet one or more of the following requirements. It shall be:

(i) **totally enclosed in arc-resistant material**

(ii) screened by arc-resistant material from materials upon which the emissions could have harmful effects

(iii) mounted so as to allow safe extinction of the emissions at a sufficient distance from materials upon which the emissions could have harmful effects

(iv) in compliance with its standard.

Arc-resistant material used for this protective measure shall be non-ignitable, of low thermal conductivity and of adequate thickness to provide mechanical stability.

21 Except where otherwise recommended by the manufacturer, spotlights and projectors rated at over 100 W and up to 300 W shall be installed at a minimum distance from combustible materials of

- a   0.5 m
- b   0.6 m
- ● c   0.8 m
- d   1.0 m.

Answer c

See Regulation 422.3.1, item (ii), which shows that option c is correct. Regulation 422.4.2 specifies similar installation requirements.

422.3.1 Except for equipment for which an appropriate product standard specifies requirements, a luminaire shall be kept at an adequate distance from combustible materials. Unless otherwise recommended by the manufacturer, a small spotlight or projector shall be installed at the following minimum distance from combustible materials:

(i)   Rating up to 100 W   0.5 m

(ii)   Over 100 and up to 300 W   0.8 m

(iii)   Over 300 and up to 500 W   1.0 m

Notes

22 In locations with increased risks of fire, motors which are automatically or remotely controlled, or which are not continuously supervised, shall be protected against excessive temperature by

○ a   a protective device that is automatically reset
◉ b   a protective device with manual reset
○ c   electronic monitoring equipment that resets
○ d   electronic monitoring equipment that restarts the motor.

Answer b

Motors which are automatically or remotely controlled are required to be protected against excessive temperature by a protective device with a manual reset (option b), otherwise the overheating will not be identified nor action taken. See Regulation 422.3.7.

422.3.7   A motor which is automatically or remotely controlled or which is not continuously supervised shall be protected against excessive temperature by a protective device with manual reset. A motor with star-delta starting shall be protected against excessive temperature in both the star and delta configurations

23 Where particular risks of fire exist, the classification for high density occupation areas with easy conditions of evacuation is

○ a   BD1
○ b   BD2
◉ c   BD3
○ d   BD4.

Answer c

See Regulation 422.2, Conditions for evacuation in an emergency, which gives the precautions where particular risks of fire exist. Note that option a, BD1, is not a relevant code.

422.2   Conditions for evacuation in an emergency
The following regulations refer to conditions:
BD2: Low density occupation, difficult conditions of evacuation
BD3: High density occupation, easy conditions of evacuation
BD4: High density occupation, difficult conditions of evacuation
(Refer to Appendix 5.)

NOTE: Authorities such as those responsible for building construction, public gatherings, fire prevention, hospitals, etc. may specify which BD condition is applicable.

24 **When considering protection against overload, the symbol for the current ensuring effective operation of the protective device in the conventional time is**

○ a I_b
○ b I_z
○ c I_n
◉ d I_2.

Answer d
Section 433 covers protection against overload current. See Regulation 433.1.1, which defines the symbol I_2 as: 'the current causing effective operation of the protective device in the conventional time'.

25 **For protection against overvoltage a 230 V electricity meter should have an impulse withstand of**

◉ a   6 kV
○ b   4 kV
○ c   2.5 kV
○ d   1.5 kV.

Answer a
A 230 V electricity meter is a category IV item and as such should have an impulse withstand of 6 kV. See Table 44.4 to identify the category and then Table 44.3 to find the impulse withstand voltage.

26 **What is the impulse category of equipment that is part of the fixed electrical installation and other equipment where a high degree of availability is expected?**

○ a   I
○ b   II
◉ c   III
○ d   IV

Answer c
Section 443 (Protection Against Overvoltages of Atmospheric Origin or due to Switching) specifies minimum impulse withstands and impulse categories and Table 44.4 provides examples of withstand categories. 'Equipment which is part of the fixed electrical installation and other equipment where a high degree of availability is expected' is an example of category III.

Table 44.4   Examples of various impulse category equipment

Category	Example
I	Equipment intended to be connected to the fixed electrical installation where protection against transient overvoltage is external to the equipment, either in the fixed installation or between the fixed installation and the equipment. Examples of equipment are household appliances, portable tools and similar loads intended to be connected to circuits in which measures have been taken to limit transient overvoltages.
II	Equipment intended to be connected to the fixed electrical installation eg household appliances, portable tools and similar loads, the protective means being either within or external to the equipment.
III	Equipment which is part of the fixed electrical installation and other equipment where a high degree of availability is expected, eg distribution boards, circuit-breakers, wiring systems, and equipment for industrial uses, stationary motors with permanent connection to the fixed installation.
IV	Equipment to be used at or in the proximity of the origin of the electrical installation upstream of the main distribution board, eg electricity meter, primary overcurrent device, ripple control unit.

From the *IEE Wiring Regulations*, Table 44.4, page 85

Section 5

27 If the construction of equipment is unsuited to the external influences of its location, it should be

- ○ a   given a plastic coating
- ○ b   given a zinc finish
- ◉ c   provided with additional protection during erection
- ○ d   supplied by SELV only.

Answer c
Regulation 512.2, External influences, provides the requirements for external influences. See Regulation 512.2.2 for specific details.

512.2.2 If the equipment does not, by its construction, have the characteristics relevant to the external influences of its location, it may nevertheless be used on condition that it is provided with appropriate additional protection in the erection of the installation. Such protection shall not adversely affect the operation of the equipment thus protected.

28 A permanent label to BS 951 bearing the words 'Safety Electrical Connection – Do Not Remove' is <u>not</u> required at

- ○ a   the connection of every earthing conductor to an earth electrode
- ○ b   the point of connection of every bonding conductor to an extraneous-conductive-part
- ○ c   the main earth terminal, where separate from the main switchgear
- ⦿ d   a main earthing bar contained within switchgear.

Answer d

You can find the requirements in Regulation group 514.13 (Warning notices: earthing and bonding connections), Regulation 514.13.1. Option d, 'a main earthing bar contained within switchgear', is not included in the list as only skilled persons should gain access, and they will know the importance of maintaining the connections with earth.

514.13.1 A permanent label to BS 951 with the words 'Safety Electrical Connection – Do Not Remove' shall be permanently fixed in a visible position at or near:
- (i)   the point of connection of every earthing conductor to an earth electrode, and
- (ii)   the point of connection of every bonding conductor to an extraneous-conductive-part, and
- (iii)   the main earth terminal, where separate from main switchgear.

29 A functional switch has to be provided for each part of the circuit

- ⦿ a   that may require independent control
- ○ b   1200 mm from the floor
- ○ c   for safe isolation
- ○ d   for emergency switching purposes.

Answer a

See Regulation 537.5.1.1, which shows that option a is the correct answer. A functional switch is neither an isolator nor an emergency switch (options c and d), and there is no requirement in BS 7671 that it has to be '1200 mm from the floor' (option b).

537.5.1.1 A functional switching device shall be provided for each part of a circuit which may require to be controlled independently of other parts of the installation.

Notes

30 A firefighter's switch shall be provided in the low voltage circuit supplying exterior electrical installations and interior discharge lighting operating at

- a   a voltage exceeding low voltage
- b   low voltage
- c   voltage band II
- d   medium voltage.

Answer a

A firefighter's switch is required in any low voltage circuit supplying exterior electrical installations operating at a voltage exceeding low voltage (that is, high voltage): see Regulation 537.6.1. This means a firefighter's switch is required for all high voltage lighting (exceeding 1000 V) including neon signs.

537.6.1 A firefighter's switch shall be provided in the low voltage circuit supplying:
(i)   exterior electrical installations operating at a voltage exceeding low voltage, and
(ii) interior discharge lighting installations operating at a voltage exceeding low voltage.

For the purpose of this regulation, an installation in a covered market, arcade or shopping mall is considered to be an exterior installation. A temporary installation in a permanent building used for exhibitions is considered not to be an exterior installation.

31 A plug and socket-outlet may be used for switching off for mechanical maintenance as long as it does not have a rating exceeding

- a   13 A
- b   16 A
- c   32 A
- d   45 A.

Answer b

The requirements for switching off for mechanical maintenance are given in Section 537. The use of a plug and socket-outlet is allowed up to 16 A. See Regulation 537.3.2.6.

Notes

537.3.2.6 A plug and socket-outlet or similar device of rating not exceeding 16 A may be used as a device for switching off for mechanical maintenance.

32 If an area within an installation undergoes a 10 °C rise in ambient temperature, the effect on the current-carrying capacity of cables will be to

- ⦿ a   decrease the value of I_z
- ○ b   increase the value of I_z
- ○ c   leave I_z unchanged
- ○ d   increase the fault current by 10 per cent.

Answer a

I_z is the current-carrying capacity of a cable for continuous service in its particular installation conditions, that is after the relevant rating factors have been applied. Appendix 4 gives guidance on cable ratings and the effect of ambient temperature: see paragraph 2.1, p253, and Tables 4B1 and 4B2. An increase in ambient temperature reduces the ambient temperature rating factor, which reduces I_z. The fault current is affected by the impedance of the cable, which is dependent upon the conductor operating temperature. An increase in the ambient temperature is likely to increase the conductor temperature, increase its impedance and so reduce the fault current.

33 A multicore 70 °C thermoplastic cable with 2.5 mm² conductors supplies a single-phase load of 20 A at 230 V a.c. over a distance of 22 metres. The voltage drop in the cable will be

- ○ a   6 V
- ○ b   6.6 V
- ⦿ c   7.92 V
- ○ d   9.2 V.

Answer c

Refer to Table 4D2B of Appendix 4. The tabulated voltage drop for 2.5 mm² cable is 18 mV/A/m.

To calculate the voltage drop in volts, the tabulated value of voltage drop (mV/A/m) has to be multiplied by the length of run in metres (L), the design current of the circuit (I_b), and divided by 1000 (to convert to volts).

$$\text{Voltage drop} = \frac{(mV/A/m) \times L \times I_b}{1000} \ V$$

So, voltage drop for the question is

$$\text{Voltage drop} = \frac{(18) \times 22 \times 20}{1000} \text{ V}$$

$$= 7.92 \text{ V}$$

34 Where a 3-core cable, with cores coloured brown, black and grey, is used as a switch wire for two-way or intermediate control, the terminations of the conductors shall be identified using

- ○ a red, blue and yellow tape
- ○ b black tape only on each core
- ⦿ c brown tape on the black and grey cores
- ○ d self-colour tape only.

Answer c

See paragraph 4 of Appendix 7 (Harmonized Cable Core Colours) or Regulation 514.4.4, which refers to Table 51. Table 51 states that line conductors in single-phase circuits shall be coloured brown.

Appendix 7
4 Intermediate and two-way switch wires in a new installation or an addition or alteration to an existing installation

Where a three-core cable with cores coloured brown, black and grey is used as a switch wire, all three conductors being line conductors, the black and grey conductors should be marked brown or L at their terminations.

35 Which one of the following cannot be used as an earth electrode?

- ○ a Earth plates
- ○ b Welded reinforcement of concrete embedded in the earth
- ○ c Earth tapes
- ⦿ d Gas and water utility pipes

Answer d

'Gas and water utility pipes' (option d) must not be used as a means of earthing, as they may be replaced by plastic pipes during the lifetime of the installation without the knowledge or understanding of the importance to the user (eg householder). See Regulation 542.2.4.

Notes

542.2.4 A metallic pipe for gases or flammable liquids shall not be used as an earth electrode. The metallic pipe of a water utility supply shall not be used as an earth electrode. Other metallic water supply pipework shall not be used as an earth electrode unless precautions are taken against its removal and it has been considered for such a use.

36 An earthing conductor buried in the ground is protected against corrosion by a sheath, but is not protected against mechanical damage. The minimum size copper conductor that may be installed is

- ○ a　　2.5 mm²
- ◉ b　　16 mm²
- ○ c　　25 mm²
- ○ d　　50 mm².

Answer b

Refer to Regulation 542.3.1 and Table 54.1 for the minimum sizes. Other regulations may require larger sizes to be used. Table 54.1 provides the minimum sizes for protection against corrosion and mechanical damage when buried.

542.3.1 Every earthing conductor shall comply with Section 543 and, where PME conditions apply, shall meet the requirements of Regulation 544.1.1 for the cross-sectional area of a main protective bonding conductor. In addition, where buried in the ground, the earthing conductor shall have a cross-sectional area not less than that stated in Table 54.1. For a tape or strip conductor, the thickness shall be such as to withstand mechanical damage and corrosion.

Table 54.1　Minimum cross-sectional area of a buried earthing conductor

	Protected against mechanical damage	Not protected against mechanical damage
Protected against corrosion by a sheath	2.5 mm² copper 10 mm² steel	**16 mm² copper** 16 mm² coated steel
Not protected against corrosion	25 mm² copper 50 mm² steel	

From the *IEE Wiring Regulations*,
Table 54.1, page 127

Notes

37 Assuming that both the line and protective conductors are of the same material, for a line conductor of 10 mm² if the protective conductor is to be selected, the minimum tabulated cross-sectional area of its associated protective conductor is

- ○ a   6 mm²
- ◉ b   10 mm²
- ○ c   16 mm²
- ○ d   35 mm².

Answer b

The minimum cross-sectional areas are set out in Section 543, Table 54.7. In this case, the size of the protective conductor is the same as the line conductor.

Table 54.7   Minimum cross-sectional area of protective conductor in relation to the cross-sectional area of associated line conductor

Cross-sectional area of line conductor S	Minimum cross-sectional area of the corresponding protective conductor	
	If the protective conductor is of the same material as the line conductor	If the protective conductor is not of the same material as the line conductor
(mm²)	(mm²)	(mm²)
$S \leq 16$	S	$\dfrac{k1}{k2} \times S$
$16 < S \leq 35$	16	$\dfrac{k1}{k2} \times 16$
$S > 35$	$\dfrac{S}{2}$	$\dfrac{k1}{k2} \times \dfrac{S}{2}$

From the *IEE Wiring Regulations*, Table 54.7, page 130

38 A radial final circuit feeding socket outlets supplying several items of data processing equipment has a total protective conductor current in normal service of 18 mA. This circuit must have a high integrity protective conductor

- ○ a   of cross-sectional area less than 1 mm²
- ◉ b   connected as a ring
- ○ c   controlled by an isolator
- ○ d   enclosed in insulated conduit only.

Answer b

This question relates to earthing requirements for the installation of equipment having high protective conductor currents (see Regulation 543.7). The answer concerns socket-outlet final circuits and is found in Regulation 543.7.2.1 (ii) (a). Options a, c and d are not requirements of the Regulations.

543.7.2.1 For a final circuit with a number of socket-outlets or connection units intended to supply two or more items of equipment, where it is known or reasonably to be expected that the total protective conductor current in normal service will exceed 10 mA, the circuit shall be provided with a high integrity protective conductor connection complying with the requirements of Regulations 543.7.1. The following arrangements of the final circuit are acceptable:

(i)   A ring final circuit with a ring protective conductor. Spurs, if provided, require high integrity protective conductor connections complying with the requirements of Regulation 543.7.1

(ii)   A radial final circuit with a single protective conductor:
   (a) the protective conductor being connected as a ring, or
   (b) a separate protective conductor being provided at the final socket-outlet by connection to the metal conduit or ducting, or
   (c) where two or more similar radial circuits supply socket-outlets in adjacent areas and are fed from the same distribution board, have identical means of short-circuit and overcurrent protection and circuit protective conductors of the same cross-sectional area, then a second protective conductor may be provided at the final socket-outlet on one circuit by connection to the protective conductor of the adjacent circuit.

(iii)   Other circuits complying with the requirements of Regulation 543.7.1.

39 A static type uninterruptible power supply source shall comply with

- ○ a   BS 3036
- ○ b   BS 1361
- ○ c   BS EN 60898
- ◉ d   BS EN 62040.

Answer d

Details of uninterruptible power supply sources (UPS) are found in Regulation 560.6.11. Options a, b and c are all protective devices.

40 Where an autotransformer is connected to a circuit having a neutral conductor, the common terminal of the winding shall be connected to the

◉ a   neutral conductor
○ b   line conductor
○ c   protective conductor
○ d   bonding conductor.

Answer a

The common terminal of an autotransformer is the terminal that is common to the input and output. Regulation 555.1.1 requires this common terminal to be connected to the neutral so that the neutral on the secondary side of the transformer is at neutral potential.

555.1.1 Where an autotransformer is connected to a circuit having a neutral conductor, the common terminal of the winding shall be connected to the neutral conductor.

Connection of common terminal of autotransformer to the neutral of the supply

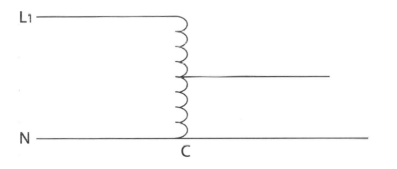

From the *Commentary on BS 7671*, Figure 13.3

41 When selecting wiring systems for safety services, the type of cable that should be used in fire conditions should comply with

○ a   BS 5467
○ b   BS 6231
○ c   BS 7211
◉ d   BS EN 50362.

Answer d

If you refer to Regulation 560.8.1, the only one of the above options listed is d. All the other options relate to cables which are not fire resistant.

Notes

560.8.1 One of the following cables shall be utilised for safety services required to operate in fire conditions:
(i)   Fire-resistant cables complying with BS EN 50362 or BS EN 50200, appropriate for the cable size, and with BS EN 60332-1-2
(ii)   Cables maintaining the necessary fire and mechanical protection.
NOTE: BS 5839-1, clause 26.2, specifies cables to BS EN 60702-1, BS 7629 and BS 7846 as suitable.

The mounting and installation of cables shall be such that the circuit integrity is maintained under fire conditions for as long as possible.

Section 6

42 When carrying out a visual inspection of an electrical installation, which one of the following does <u>not</u> have to be verified?

○ a   The methods of protection against electric shock
◉ b   The electricity supplier
○ c   The connection of conductors
○ d   The presence of undervoltage protective devices

Answer b
Knowledge of the electricity supplier (option b) is not necessary when carrying out a visual inspection. Regulation 611.3 lists the inspection items, which include options a, c and d.

43 Certain information must be made available to persons carrying out inspection and testing of an installation before the testing commences. One such item of information would be

○ a   the name of the client
○ b   the name of the person who designed the installation
○ c   the length of cable runs in the installation
◉ d   any circuit or equipment vulnerable to a typical test.

Answer d
It is necessary to know which circuits and equipment might be vulnerable to the tests of Chapter 61 (option d) before testing commences to avoid causing damage. Refer to Regulations 610.2 and 514.9.1.

610.2 The result of the assessment of the fundamental principles, Section 131, the general characteristics required by Sections 311 to 313, together with the information required by Regulation 514.9.1, shall be made available to the person or persons carrying out the inspection and testing.

514.9.1 A legible diagram, chart or table or equivalent form of information shall be provided indicating in particular:
(i)   the type and composition of each circuit (points of utilisation served, number and size of conductors, type of wiring), and
(ii)   the method used for compliance with Regulation 410.3.2, and
(iii) the information necessary for the identification of each device performing the functions of protection, isolation and switching, and its location, and
(iv) any circuit or equipment vulnerable to a typical test.

44 When carrying out an inspection of a new installation, it is _not_ necessary to verify the

- ◉ a   total earth fault loop impedance for each circuit
- ○ b   connection of conductors
- ○ c   methods of protection against electric shock
- ○ d   presence of diagrams, instructions and similar information.

Answer a
This question relates to inspection, not testing, so option a is the only option that it is not necessary to verify. Options b, c and d are part of the inspection check list given in Regulation 611.3.

45 An insulation resistance test is to be carried out on a 3-phase 400 V circuit. The test voltage and minimum acceptable reading would be

- ○ a   250 V a.c. and 0.5 MΩ
- ○ b   500 V d.c. and 0.5 MΩ
- ◉ c   500 V d.c. and 1 MΩ
- ○ d   800 V d.c. and 0.5 Ω.

Answer c
Insulation resistance is measured with d.c. test voltages to eliminate the effects of capacitance and inductance, so option a is incorrect. Minimum insulation resistance requirements for testing are given in Regulation 612.3.2 and Table 61.

Notes

From the *IEE Wiring Regulations*,
Table 61, page 158

Table 61   Minimum values of insulation resistance

Circuit nominal voltage (V)	Test voltage d.c. (V)	Minimum insulation resistance (MΩ)
SELV and PELV	250	≥ 0.5
Up to and including 500 V with the exception of the above systems	**500**	**≥ 1.0**
Above 500 V	1000	≥ 1.0

46 When an addition is made to an existing installation, the contractor shall record on the Electrical Installation Certificate or the Minor Electrical Installation Works Certificate any

○ a   changes in ownership
○ b   records of repair over the last five years
◉ c   defects in the existing installation
○ d   voltage drop in the longest circuit.

Answer c

It is important that any identified 'defects in the existing installation' are recorded on the certificates (option c). Defects affecting the safety of the addition must be rectified. Refer to Regulation 633.2.

633.2   The contractor or other person responsible for the new work, or a person authorized to act on their behalf, shall record on the Electrical Installation Certificate or the Minor Electrical Installation Works Certificate, any defects found, so far as is reasonably practicable, in the existing installation.

47 After completion of a periodic inspection, the completed documentation shall be given to the

◉ a   person ordering the inspection
○ b   local authority
○ c   insurance company
○ d   main contractor.

Answer a

After completion of a periodic inspection, the completed documentation is given to the person ordering the inspection. See Regulation 634.1.

634.1 Following the periodic inspection and testing described in Chapter 62, a Periodic Inspection Report, together with a schedule of inspections and a schedule of test results, shall be given by the person carrying out the inspection, or a person authorized to act on their behalf, to the person ordering the inspection. These schedules shall be based on the models given in Appendix 6. The schedule of test results shall record the results of the appropriate tests required of Chapter 61.

Section 7

48 In a room containing a bath, electrical equipment installed in zone 0 shall have a degree of protection of at least

- ○ a   IPX5
- ○ b   IP5X
- ○ c   IP7X
- ◉ d   IPX7.

Answer d

The requirements for external influences in a bathroom are given in Regulation 701.512.2. For zone 0, the zone which is likely to be totally immersed in water, the requirement is IPX7.

701.512.2 External influences

Installed electrical equipment shall have at least the following degrees of protection:
(i)   In zone 0: IPX7
(ii)   In zones 1 and 2: IPX4.

49 A flush downlighter in the ceiling less than 2.5 m height above floor level over a bath will need to fulfil the requirements of

- ○ a   zone 2
- ◉ b   zone 1
- ○ c   zone 0
- ○ d   all zones.

Answer b

See Regulation 701.32.1. As the area above a bath is zone 1, then the requirements of zone 1 must apply in this case.

702.32.1 For electrical equipment in parts of walls or ceilings limiting the zones specified in Regulations 701.32.2 to 701.32.4, but being part of the surface of that wall or ceiling, the requirements for the respective zone apply.

Notes

50 Which of the following protective measures is permitted in a room containing a bath or shower?

- ○ a   Obstacles
- ○ b   Placing out of reach
- ◉ c   Automatic disconnection of supply
- ○ d   Non-conducting locations

Answer c

Options a, b and d are not permitted as a protective measure, so option c is correct. See Regulations 701.410.3.5 and 701.410.3.6.

51 In a swimming pool or other basin, the metallic covering or sheath of a wiring system in zones 0, 1 or 2 shall

- ○ a   not be used
- ○ b   not be earthed
- ◉ c   be connected to the supplementary bonding
- ○ d   have reinforced insulation.

Answer c

There are clear requirements for wiring systems in swimming pools in Regulation 702.52 and in particular Regulation 702.522.21.

702.522.21   Erection according to the zones

In zones 0, 1 and 2, any metallic sheath or metallic covering of a wiring system shall be connected to the supplementary equipotential bonding.

NOTE: Cables should preferably be installed in conduits made of insulating material.

52 The requirements of Section 704 of BS 7671 apply to

- ○ a   cloakrooms
- ○ b   offices
- ◉ c   construction and demolition site installations
- ○ d   toilets.

Answer c

There are no additional requirements for the cloakrooms, offices and toilets of construction sites in Section 704. They are not considered part of the building construction site and this is made clear by Regulation 704.1.1. Options a, b and d are excluded by item (vii).

704.1.1 The particular requirements of this section apply to temporary installations for construction and demolition sites during the period of the construction or demolition work, including, for example, the following:

(i) Construction work of new buildings
(ii) Repair, alteration, extension, demolition of existing buildings or parts of existing buildings
(iii) Engineering works
(iv) Earthworks
(v) Work of similar nature.

The requirements apply to fixed or movable installations.

The regulations do not apply to:

(vi) installations covered by the IEC 60621 series 2, where equipment of a similar nature to that used in surface mining applications is involved
(vii) installations in administrative locations of construction sites (eg offices, cloakrooms, meeting rooms, canteens, restaurants, dormitories, toilets), where the general requirements of Part 1 to 6 apply.

53 In locations where livestock is kept, for all circuits other than socket-outlet circuits, an RCD shall be provided with a rating not exceeding

○ a   500 mA
◉ b   300 mA
○ c   100 mA
○ d   30 mA.

Answer b

A rated residual operating current $I_{\Delta n}$ not exceeding 300 mA is given in Regulation 705.411.1, item (iii).

Notes

705.411.1 General

In circuits, whatever the type of earthing system, the following disconnection device shall be provided:

(i) In final circuits supplying socket-outlets with rated current not exceeding 32 A, an RCD with a rated residual operating current not exceeding 30 mA

(ii) In final circuits supplying socket-outlets with rated current more than 32 A, an RCD with a rated residual operating current not exceeding 100 mA

(iii) **In all other circuits, RCDs with a rated residual operating current not exceeding 300 mA.**

---

54 Equipment on a pontoon in a marina that is subject to impact to level AG2 should have a mechanical protection code of

- ○ a IPX4
- ○ b IPX8
- ○ c IP55
- ◉ d IK08.

Answer d

Regulation 709.512.2.1.4 details the requirement for impact protection. See item (iii) which shows option d to be the answer. Options a, b and c are codes for water and foreign solid bodies ingress.

709.512.2.1.4 Impact (AG)

Equipment installed on or above a jetty, wharf, pier or pontoon shall be protected against mechanical damage (Impact of medium severity AG2). Protection shall be afforded by one or more of the following:

(i) The position or location selected to avoid being damaged by any reasonably foreseeable impact

(ii) The provision of local or general mechanical protection

(iii) **Installing equipment complying with a minimum degree of protection for external mechanical impact IK08 (see BS EN 62262).**

---

55 For marinas, the classification of external influence which does not need to be considered is

- ○ a AD
- ○ b AE
- ○ c AF
- ◉ d AP.

Answer d

Influence AP is seismic and not a requirement of Regulations 709.512.2.1.1 to 709.512.2.1.4, which between them quote options a, b and c together with AG.

56 In caravans, each final circuit shall be protected against overcurrent by a device that disconnects

- ◉ a all live conductors in that circuit
- ○ b the line conductors only
- ○ c line, neutral and protective conductors
- ○ d the caravan and site supply system.

Answer a

As the polarity of caravan site supplies is not specified in all countries, caravan vehicle wiring must be suitable for unpolarised supplies, ie any overcurrent devices must disconnect all live conductors. See Regulation 721.43.1.

721.43.1 Final circuits

Each final circuit shall be protected by an overcurrent protective device which disconnects all live conductors of that circuit.

57 Electric dodgems shall only be operated at voltages not exceeding

- ○ a 120 V a.c. or 50 V d.c.
- ◉ b 50 V a.c. or 120 V d.c.
- ○ c 110 V a.c. reduced low voltage
- ○ d 1000 V a.c. or 1500 V d.c.

Answer b

See Regulation 740.55.9.

740.55.9 Electric dodgems

Electric dodgems shall only be operated at voltages not exceeding 50 V a.c. or 120 V d.c. The circuit shall be electrically separated from the supply mains by means of a transformer in accordance with BS EN 61558-2-4 or a motor-generator set.

Section 8

58 The requirements of the licensing authority should be ascertained for

- ◉ a   installations on caravan parks
- ○ b   temporary supplies
- ○ c   types of earthing system to be used
- ○ d   the design of the installation.

Answer a

For installations on caravan parks, the requirements of the licensing authority need to be ascertained before carrying out designs or testing installations. Certificates indicating compliance with BS 7671 may be required. Local authorities may have particular requirements on frequency of testing. Refer to paragraph 6 of Appendix 2 of BS 7671.

Appendix 2

6 For installations in theatres and other places of public entertainment, and on caravan parks, the requirements of the licensing authority should be ascertained. Model Standards were issued by the Department of the Environment in 1977 under the Caravan Sites and Control of Development Act 1960 as guidance for local authorities.

59 I_2 can be greater than $1.45I_z$ when the overcurrent device

- ○ a   is providing overload protection
- ○ b   is a circuit-breaker
- ○ c   is rated at over 100 A
- ◉ d   is providing fault current protection only.

Answer d

I_2 can be greater than $1.45I_z$ when the overcurrent device 'is providing fault current protection only' (option d). When overload protection is being provided I_2 must be equal to or less than $1.45I_z$. See Section 3 of Appendix 4 and Regulation 434.5.2.

Appendix 4, Section 3, end paragraph

Where the overcurrent device is intended to afford fault current protection only, I_n can be greater than I_z and I_2 can be greater than $1.45 I_z$. The protective device must be selected for compliance with Regulation 434.5.2.

60 External influences coded BE are classified as

- ⦿ a   nature of processed or stored materials
- ○ b   conditions of evacuation in an emergency
- ○ c   movement of air
- ○ d   capability of persons.

Answer a

Appendix 5 (Classification of External Influences) provides the answer.

Nature of processed or stored materials

Code	Class designation	Characteristics	Applications and examples
BE1	No significant risk		
BE2	Fire risks	Manufacture, processing or storage of flammable materials including presence of dust	Barns, woodworking shops, paper factories
BE3	Explosion risks	Processing or storage of explosive or low flash-point materials including presence of explosive dusts	Oil refineries, hydrocarbon stores
BE4	Contamination risks	Presence of unprotected foodstuffs, pharmaceutics, and similar products without protection	Foodstuff industries, kitchens. Certain precautions may be necessary, in the event of fault, to prevent processed materials being contaminated by electrical equipment, eg by broken lamps

From the *IEE Wiring Regulations*, extract from Appendix 5, page 330

Notes

Answer key

Sample test 1

Question	Answer	Question	Answer
1	d	31	b
2	d	32	a
3	c	33	c
4	a	34	c
5	b	35	d
6	a	36	b
7	c	37	b
8	c	38	b
9	c	39	d
10	a	40	a
11	c	41	d
12	d	42	b
13	d	43	d
14	a	44	a
15	a	45	c
16	b	46	c
17	c	47	a
18	c	48	d
19	d	49	b
20	a	50	c
21	c	51	c
22	b	52	c
23	c	53	b
24	d	54	d
25	a	55	d
26	c	56	a
27	c	57	b
28	d	58	a
29	a	59	d
30	a	60	a

Exam practice 2

Exam practice 2

Notes

Sample test 2

The sample test below has 60 questions, the same number as the online exam, and its structure follows that of the online exam. The test appears first without answers, so you can use it as a mock exam. It is then repeated with worked-through answers and extracts from the *IEE Wiring Regulations*. Finally, there is an answer key for easy reference.

Answer the questions by filling in the circle next to your chosen option.

Section 1

1 BS 7671 includes the requirements for

○ a   fixed wiring for information and communication technology, signalling and control
○ b   radio interference suppression equipment
○ c   systems for the distribution of electricity to the public
○ d   equipment on board ships and aircraft.

2 BS 7671 requires protection against electric shock to be provided by basic and fault protection. One method common to both is

○ a   the use of RCDs
○ b   the use of Class II equipment
○ c   preventing current from passing through any person or livestock
○ d   the use and implementation of equipotential bonding.

3 Which one of the following types of electrical installation is <u>not</u> covered by BS 7671?

○ a   High protective conductor current installations
○ b   Lightning protection of buildings
○ c   Conducting locations with restricted movement
○ d   Highway power supplies

4 The selection of the type of wiring and method of installation is <u>not</u> influenced by

○ a   the nature of the location
○ b   the load current
○ c   the value of the prospective short-circuit current
○ d   the nature of the structure supporting the wiring.

Section 2

5   Band II voltage exceeds

○ a   2500 V d.c. between live conductors
○ b   2500 V d.c. between conductors and Earth
○ c   1000 V a.c. rms
○ d   120 V d.c. ripple-free.

6   A protective conductor connecting exposed-conductive-parts of equipment to the main earthing terminal is known as the

○ a   earthing conductor
○ b   main bonding conductor
○ c   circuit protective conductor
○ d   bonding conductor.

7   An area or temporary structure used for display, marketing or sales is defined as

○ a   a booth
○ b   a stand
○ c   an exhibition
○ d   a show.

Section 3

8   To ensure continuity of supply to one part of the installation in the event of failure of another part, one approach is

○ a   for a separate circuit to be installed for each part
○ b   to use protective equipotential bonding
○ c   to install larger fuses
○ d   to install a 100 mA S type RCD.

9   An assessment must be made of the harmful effects of harmonic currents of equipment on other electrical equipment. This assessment is a check of

○ a   external influences
○ b   compatibility
○ c   diversity
○ d   maintainability.

Notes

10 Every installation must be divided into circuits as necessary in order to

○ a   reduce the cost of installation
○ b   make installation easier
○ c   install a cooker
○ d   facilitate safe inspection, testing and maintenance.

11 Which one of the following is <u>not</u> a potential characteristic of equipment to be considered for harmful effects on other equipment?

○ a   Transient overvoltages
○ b   d.c. feedback
○ c   Prospective short-circuit current
○ d   Excessive protective conductor current

12 When making an assessment of the frequency and quality of maintenance, a factor to be considered is that

○ a   power factor is monitored
○ b   protective measures for safety remain effective
○ c   starting currents are at a minimum
○ d   unbalanced loads need to be checked more frequently.

Section 4

13 A BS 88-6 32 A fuse with U_0 of 230 V has a maximum tabulated Z_s for 5 seconds disconnection of

○ a $0.36\,\Omega$
○ b $1.04\,\Omega$
○ c $1.44\,\Omega$
○ d $1.84\,\Omega$.

14 Which of the following is a requirement in terms of fault protection where the protective measure of electrical separation is employed?

○ a   The voltage of the separated circuit is not more than 500 V
○ b   Flexible cables and cords liable to mechanical damage are concealed
○ c   Live parts are connected to a protective conductor
○ d   Exposed-conductive-parts are connected to Earth

15 An RCD can be used as additional protection in the event of failure of the provision for basic protection when the RCD has a rating not exceeding

- a   30 mA
- b   100 mA
- c   300 mA
- d   500 mA.

16 Basic protection can be provided by

- a   protective earthing
- b   barriers
- c   automatic disconnection in case of a fault
- d   functional earth.

17 An obstacle is <u>not</u> intended to

- a   prevent unintentional bodily approach to live parts
- b   prevent intentional contact with live parts during operation
- c   be removed by use of a key or tool
- d   be secured so as to prevent unintentional removal.

18 For a non-conducting location, the resistance of an insulating floor or wall at every point of measurement specified in Part 6 shall not be less than

- a   45 kΩ where the voltage to earth does not exceed 230 V
- b   50 kΩ where the voltage to earth does not exceed 500 V
- c   55 kΩ where the voltage to earth does not exceed 750 V
- d   60 kΩ where the voltage to earth does not exceed 1000 V.

19 Where electrical separation is used to supply more than one item of current-using equipment

- a   warning notices may be omitted
- b   the length of the wiring system should not exceed 500 m
- c   socket-outlets should not have a protective conductor contact
- d   the provisions for basic protection are not required.

20 In locations with risks of fire due to the nature of processed or stored materials, enclosures such as heaters and resistors shall not attain in normal conditions higher surface temperatures than

- ○ a 90 °C
- ○ b 85 °C
- ○ c 60 °C
- ○ d 55 °C.

21 In a paper mill installation wired in PVC single-core cables in conduit and trunking, where a fire may be caused by a resistive fault in overhead heaters with film elements, protection against an insulation fault shall be provided by a

- ○ a BS 88 fuse
- ○ b BS EN 60898 cb
- ○ c 300 mA RCD
- ○ d 30 mA RCD.

22 The maximum temperature for the exposed metallic casing of a wall-mounted, cord-operated infra-red heater is

- ○ a 55 °C
- ○ b 65 °C
- ○ c 70 °C
- ○ d 80 °C.

23 Where fault current protection is <u>not</u> required, protection against overload current is generally provided by

- ○ a BS EN 60898 fuses
- ○ b an inverse-time-lag protective device
- ○ c a BS 3036 semi-enclosed fuse
- ○ d RCDs.

24 Regulations 434.2 and 434.2.1 do <u>not</u> allow the fault current protective device to be placed on the load side of the reduction in current-carrying capacity of a conductor if

○ a   the length of conductor is less than 3 m
○ b   the length of conductor exceeds 3 m
○ c   it is erected in such a manner as to reduce to a minimum the risk of fault current
○ d   it is erected in such a manner as to reduce to a minimum the risk of fire or danger to persons.

25 In the formula

$$t = \frac{k^2 S^2}{I^2}$$

what is the k factor for copper conductors insulated with 70 °C thermoplastic and an assumed initial temperature of 70 °C?

○ a   100/86
○ b   115/103
○ c   141
○ d   143

26 What is the required minimum impulse withstand voltage for category IV equipment for use in an installation with nominal voltage of 400/690?

○ a   2.5 kV
○ b   4 kV
○ c   6 kV
○ d   8 kV

Section 5

27 When selecting equipment for an installation, the person who makes the final decision over its safety would be the

○ a   representative of a regional supply company
○ b   representative of a national inspection body
○ c   designer or other person responsible
○ d   inspector and tester.

Notes

Notes

28 Unless otherwise confirmed as suitable, switchgear, protective devices and accessories shall not be connected to conductors intended to operate at a temperature in excess of

- ○ a   30 °C
- ○ b   50 °C
- ○ c   70 °C
- ○ d   90 °C.

29 A 70 °C thermoplastic insulated and sheathed cable to BS 6004 is installed in an ambient temperature of 25 °C for part of the run and then enters an area at 40 °C. The effect on the cable will be to

- ○ a   require clips at greater intervals
- ○ b   reduce the voltage drop
- ○ c   increase the bending radius
- ○ d   decrease its current-carrying capacity.

30 A single-phase load of 13 A is supplied via a 15 A BS 1361 fuse using single-core 70 °C thermoplastic copper cables installed to method 4 (Reference Method B). The rating factor for grouping is 0.7 and for ambient temperature 0.87, and overload protection is to be provided. The minimum acceptable size cable would be

- ○ a   1 mm²
- ○ b   1.5 mm²
- ○ c   2.5 mm²
- ○ d   4 mm².

31 The minimum information contained on a periodic inspection and testing notice would be the

- ○ a   date of inspection and inspector's name
- ○ b   date of next inspection and client's name
- ○ c   date of last inspection and date of next inspection
- ○ d   company's address and date of next inspection.

32 When conductors are identified by numbers, the neutral shall be identified by the number

- ○ a   0
- ○ b   4
- ○ c   5
- ○ d   14.

33 Which one of the following devices would <u>not</u> be suitable to provide overcurrent protection?

- ○ a   Residual current device to BS 4293
- ○ b   Cartridge fuse to BS 1362
- ○ c   Cartridge fuse to BS 88-6
- ○ d   Rewirable fuse to BS 3036

34 Under fault conditions, cables are subjected to stress and thermal effects. One item <u>not</u> normally encountered would be

- ○ a   electrochemical effects
- ○ b   electromechanical stresses
- ○ c   electromagnetic effects
- ○ d   thermal damage.

35 Every connection between conductors, or between conductors and equipment, need <u>not</u> be selected to account for the

- ○ a   conductor length
- ○ b   material of the conductor
- ○ c   number and shape of the wires forming the conductor
- ○ d   cross-sectional area of the conductor.

36 Where more than one firefighter's switch is installed on any one building, each switch must be

- ○ a   not more than 3.75 m from the ground
- ○ b   in a locked location to prevent nuisance operation
- ○ c   electrically linked
- ○ d   clearly marked.

37 Where safety depends upon the direction of rotation of a motor, provision shall be made to

- ○ a   measure the maximum and minimum speed
- ○ b   monitor the slip frequency
- ○ c   prevent reverse operation
- ○ d   prevent operation of overload devices.

Notes

38 A 16 mm² multicore cable incorporating a copper protective conductor is insulated with 90 °C thermosetting insulation. Given a fault current of 500 A and a disconnection time of 0.4 second, the minimum acceptable size of cpc would be

○ a   1.5 mm²
○ b   2.5 mm²
○ c   4.0 mm²
○ d   6.0 mm².

39 When using the alternative method to Regulation 543.1.3 of sizing a copper protective conductor, the minimum size to be used when a copper line conductor has a csa of 50 mm² would be

○ a   16 mm²
○ b   25 mm²
○ c   35 mm²
○ d   50 mm².

40 The maximum current rating of a protective device protecting a circuit including B22 bayonet cap lampholders would be

○ a   5 A
○ b   6 A
○ c   16 A
○ d   20 A.

41 Which of the following is <u>not</u> a BS 7671 requirement when installing a heating cable laid directly in soil?

○ a   It is completely embedded in the soil
○ b   It does not suffer damage in the event of normal movement
○ c   It complies with the manufacturer's instructions
○ d   It is protected by a 500 mA RCD

Section 6

42 During the installation inspection process, which one of the following need <u>not</u> be checked?

○ a   The external earth fault loop impedance
○ b   Equipment in compliance with Section 511
○ c   Correct selection and erection
○ d   Visible damage or defects to installed equipment

43 When testing SELV circuits, the minimum acceptable insulation resistance value is

- ○ a   0.25 MΩ
- ○ b   0.5 MΩ
- ○ c   1 MΩ
- ○ d   2 MΩ.

44 Defects identified during an initial verification should be

- ○ a   noted on the certificate, and the client should be informed
- ○ b   pointed out to the main contractor when the certificate has been completed
- ○ c   made good before the certificate is issued
- ○ d   put on a snagging list to be rectified later.

45 Which of the following is not a requirement for periodic inspection and testing?

- ○ a   Ensuring the safety of persons and livestock from shock and burns
- ○ b   Ensuring the property is protected against damage from fire and heat due to an installation fault
- ○ c   Confirmation that the installation is not damaged as to impair safety
- ○ d   Confirmation that the supply to the installation meets all the requirements of the Electricity Safety, Quality and Continuity Regulations

46 Additional work is carried out to an installation, which comprises a socket-outlet added to an existing circuit. The paperwork to be completed would be

- ○ a   a Periodic Inspection Report
- ○ b   a Schedule of Test Results
- ○ c   a Minor Electrical Installation Works Certificate
- ○ d   a Schedule of Inspections.

47 When completing an Electrical Installation Certificate, the person who does not have to sign the certificate would be the

- ○ a   tester
- ○ b   client
- ○ c   constructor
- ○ d   designer.

Notes

Section 7

48 For an electric floor heating system in a bathroom, a fine mesh metallic grid is <u>not</u> required to be connected to the protective conductor of the supply circuit if the protective measure used is

- ○ a   SELV
- ○ b   electrical separation
- ○ c   a non-conducting location
- ○ d   an earth-free equipotential zone.

49 Access to the space under a bath is gained by unscrewing the bath panel retaining clips. This space is classified as

- ○ a   zone 0
- ○ b   zone 1
- ○ c   outside the zones
- ○ d   zone 2.

50 For SELV and PELV circuits in a bathroom, basic protection for equipment in all zones shall be provided by

- ○ a   obstacles
- ○ b   barriers or enclosures
- ○ c   non-conducting location
- ○ d   placing out of reach.

51 Electrical equipment in zones 0 and 1 of a fountain shall have mechanical protection to a minimum classification of

- ○ a   AG0
- ○ b   AG1
- ○ c   AG2
- ○ d   AG3.

52 Which of the following would <u>not</u> be a requirement for an Assembly for Construction Sites (ACS), supplying mobile current-using equipment on a construction site?

- ○ a   Main equipotential bonding
- ○ b   Overcurrent protective devices
- ○ c   Devices for fault protection
- ○ d   Socket-outlets

53 Which one of the following is <u>not</u> included in the scope of Section 705?

- ○ a   Chicken-houses
- ○ b   Stables
- ○ c   Piggeries
- ○ d   Staff residences

54 In conducting locations with restricted movement, supplies to handlamps shall be protected by

- ○ a   SELV
- ○ b   reduced low voltage supplies
- ○ c   RCD or RCBO
- ○ d   PELV.

55 The minimum height of overhead conductors on caravan sites, when installed in vehicle movement areas, is

- ○ a   3.0 m
- ○ b   3.5 m
- ○ c   5.8 m
- ○ d   6.0 m.

56 The particular requirements for mobile or transportable units apply to

- ○ a   generating sets
- ○ b   mobile workshops
- ○ c   mobile machinery to BS EN 60204-1
- ○ d   traction equipment of electric vehicles.

57 Electrical equipment in a circus installation must have a degree of protection of at least

- ○ a   IP33
- ○ b   IP4X
- ○ c   IP44
- ○ d   IPX8.

Notes

Section 8

58 The ambient air temperature rating factor for 90 °C thermosetting cables operating in an ambient air temperature of 60 °C is

- ○ a   0.50
- ○ b   0.56
- ○ c   0.63
- ○ d   0.71.

59 A BS 3036 100 A fuse, when carrying 1800 A, has a disconnection time of

- ○ a   0.2 second
- ○ b   0.4 second
- ○ c   4 seconds
- ○ d   5 seconds.

60 A 2.5 mm² thermoplastic insulated and sheathed flat cable with protective conductor is laid in a ceiling beneath thermal insulation 80 mm thick in contact with the ceiling board, as shown in the figure below. What is its installed rating?

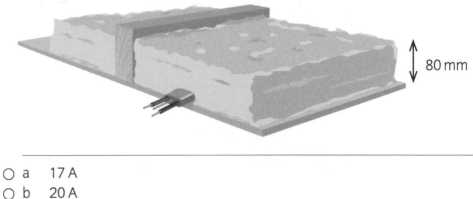

- ○ a   17 A
- ○ b   20 A
- ○ c   21 A
- ○ d   27 A

Questions and answers

The questions in sample test 2 are repeated below with worked through answers and extracts from the *IEE Wiring Regulations* where appropriate.

Section 1

1 BS 7671 includes the requirements for

◉ a   fixed wiring for information and communication technology, signalling and control
○ b   radio interference suppression equipment
○ c   systems for the distribution of electricity to the public
○ d   equipment on board ships and aircraft.

Answer a
See Part 1, Chapter 11, Scope. The installations covered by BS 7671 are listed in Regulation 110.1 and item (xix) is 'fixed wiring for information and communication technology, signalling, control and the like' (option a). Exclusions from the scope of BS 7671 (including options b, c and d) can be found in Regulation 110.2.

2 BS 7671 requires protection against electric shock to be provided by basic and fault protection. One method common to both Is

○ a   the use of RCDs
○ b   the use of Class II equipment
◉ c   preventing current from passing through any person or livestock
○ d   the use and implementation of equipotential bonding.

Answer c
Refer to Chapter 13 (Fundamental Principles), Regulations 131.2.1 and 131.2.2. Preventing a current from passing through the body of any person or livestock (option c) is a method common to both basic protection and fault protection.

131.2.1 Basic protection (protection against direct contact)
NOTE: For low voltage installations, systems and equipment, 'basic protection' generally corresponds to protection against 'direct contact'.

Persons and livestock shall be protected against dangers that may arise from contact with live parts of the installation.

This protection can be achieved by one of the following methods:
(i) **Preventing a current from passing through the body of any person or any livestock**
(ii) Limiting the current which can pass through a body to a non-hazardous value.

131.2.2 Fault protection (protection against indirect contact)
NOTE: For low voltage installations, systems and equipment, 'fault protection' generally corresponds to protection against 'indirect contact', mainly with regard to failure of basic insulation.

Persons and livestock shall be protected against dangers that may arise from contact with exposed-conductive-parts during a fault.

This protection can be achieved by one of the following methods:
(i) **Preventing a current resulting from a fault from passing through the body of any person or any livestock**
(ii) Limiting the magnitude of a current resulting from a fault, which can pass through a body, to a non-hazardous value
(iii) Limiting the duration of a current resulting from a fault, which can pass through a body, to a non-hazardous time period.

In connection with fault protection, the application of the method of protective equipotential bonding is one of the important principles for safety.

3 **Which one of the following types of electrical installation is <u>not</u> covered by BS 7671?**

○ a   High protective conductor current installations
◉ b   Lightning protection of buildings
○ c   Conducting locations with restricted movement
○ d   Highway power supplies

Answer b
The requirements for 'Lightning protection of buildings' (option b) are not covered by BS 7671; they can be found in BS EN 62305. See Regulation 110.2.

110.2 Exclusions from scope
The Regulations do not apply to the following installations:
(i)   Systems for the distribution of electricity to the public
(ii)   Railway traction equipment, rolling stock and signalling equipment
(iii) Equipment of motor vehicles, except those to which the requirements of the Regulations concerning caravans or mobile units are applicable

(iv)   Equipment on board ships covered by BS 8450
(v)   Equipment of mobile and fixed offshore installations
(vi)   Equipment of aircraft
(vii)   Those aspects of mines and quarries specifically covered by
        Statutory Regulations
(viii) Radio interference suppression equipment, except so far as it affects
        safety of the electrical installation
**(ix)   Lightning protection systems for buildings and structures
        covered by BS EN 62305**
(x)   Those aspects of lift installations covered by relevant parts of
        BS 5655 and BS EN 81-1
(xi)   Electrical equipment of machines covered by BS EN 60204.

4   **The selection of the type of wiring and method of installation is
    <u>not</u> influenced by**

○ a   the nature of the location
◉ b   the load current
○ c   the value of the prospective short-circuit current
○ d   the nature of the structure supporting the wiring.

Answer b
The nature of the location (option a), the value of the prospective short-circuit current (option c) and the nature of the structure supporting the wiring (option d) affect decisions on the type and method of wiring, for the wiring has to be suitable for the location, for fixing to the structure and robust enough to withstand fault currents. Option b, the load current, is not an influence. See Regulation 132.7.

132.7 Type of wiring and method of installation
The choice of the type of wiring system and the method of installation shall include consideration of the following:
(i)   The nature of the location
(ii)   The nature of the structure supporting the wiring
(iii)   Accessibility of wiring to persons and livestock
(iv)   Voltage
(v)   The electromechanical stresses likely to occur due to short-circuit
        and earth fault currents
(vi)   Electromagnetic interference
(vii)   Other external influences (eg mechanical, thermal and those
        associated with fire) to which the wiring is likely to be exposed
        during the erection of the electrical installation or in service.

Notes

Section 2

5 Band II voltage exceeds

- ○ a 2500 V d.c. between live conductors
- ○ b 2500 V d.c. between conductors and Earth
- ○ c 1000 V a.c. rms
- ◉ d 120 V d.c. ripple-free.

Answer d

Part 2, Definitions, states under 'Voltage bands' that low voltage will normally fall within voltage Band II and that Band II voltages do not exceed 100 V a.c. Rms or 1500 V d.c. Part 2 also defines low voltage as exceeding extra-low voltage but not exceeding 100 V a.c. or 1500 V d.c.

Voltage bands

Band I

Band I covers:
- installations where protection against electric shock is provided under certain conditions by the value of voltage;
- installations where the voltage is limited for operational reasons (eg telecommunications, signalling, bell, control and alarm installations).

Extra-low voltage (ELV) will normally fall within voltage Band I.

Band II

Band II contains the voltages for supplies to household and most commercial and industrial installations.

Low voltage (LV) will normally fall within voltage Band II.

NOTE: Band II voltages do not exceed 1000 V a.c. rms or 1500 V d.c.

6 A protective conductor connecting exposed-conductive-parts of equipment to the main earthing terminal is known as the

- ○ a earthing conductor
- ○ b main bonding conductor
- ◉ c circuit protective conductor
- ○ d bonding conductor.

Answer c

See Part 2 Definitions, Circuit protective conductor.

Circuit protective conductor (cpc). A protective conductor connecting exposed-conductive-parts of equipment to the main earthing terminal.

Earthing and protective conductor terms

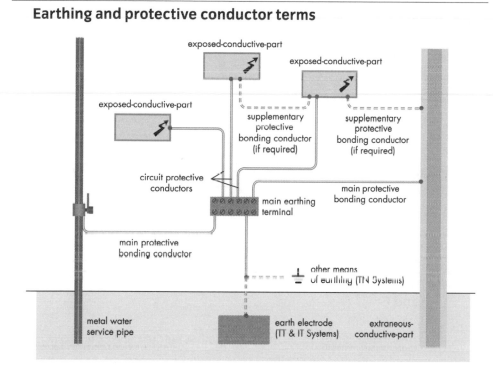

7 **An area or temporary structure used for display, marketing or sales is defined as**

○ a a booth
◉ b a stand
○ c an exhibition
○ d a show.

Answer b

See Part 2 Definitions. Option b is clearly defined as an 'area or temporary structure used for display, marketing or sales'.

Section 3

8 **To ensure continuity of supply to one part of the installation in the event of failure of another part, one approach is**

◉ a for a separate circuit to be installed for each part
○ b to use protective equipotential bonding
○ c to install larger fuses
○ d to install a 100 mA S type RCD.

Notes

Answer a

See Regulation 314.2, which shows that the answer to the question is option a.

314.2 Separate circuits shall be provided for parts of the installation which need to be separately controlled, in such a way that those circuits are not affected by the failure of other circuits, and due account shall be taken of the consequences of the operation of any single protective device.

9 **An assessment must be made of the harmful effects of harmonic currents of equipment on other electrical equipment. This assessment is a check of**

- ○ a   external influences
- ◉ b   compatibility
- ○ c   diversity
- ○ d   maintainability.

Answer b

One item of equipment may detrimentally affect other items, but if an item of equipment can operate satisfactorily with another, it is said to be compatible. See Chapter 33 (Compatibility) and Regulation 331.1.

331.1 An assessment shall be made of any characteristics of equipment likely to have harmful effects upon other electrical equipment or other services or likely to impair the supply, for example, for co-ordination with concerned parties eg petrol stations, kiosks and shops within shops. Those characteristics include, for example:
(i)   transient overvoltages
(ii)   undervoltage
(iii)   unbalanced loads
(iv)   rapidly fluctuating loads
(v)   starting currents
(vi)   harmonic currents
(vii) leakage currents
(viii) excessive protective conductor current
(ix)   d.c. feedback
(x)   high-frequency oscillations
(xi)   necessity for additional connections to Earth
(xii) power factor.

For an external source of energy the distributor shall be consulted regarding any equipment of the installation having a characteristic likely to have significant influence on the supply.

10 Every installation must be divided into circuits as necessary in order to

- ○ a   reduce the cost of installation
- ○ b   make installation easier
- ○ c   install a cooker
- ◉ d   facilitate safe inspection, testing and maintenance.

Answer d

The purpose of BS 7671 is to provide for safety, including safe operation and facilities for inspection and testing. See Regulation 314.1, which shows that even though options a, b and c might be true in particular circumstances and so are logical, they are not the mandatory reason.

314.1 Every installation shall be divided into circuits, as necessary, to

(i)   avoid hazards and minimize inconvenience in the event of a fault

(ii)   facilitate safe inspection, testing and maintenance (see also Section 537)

(iii) take account of danger that may arise from the failure of a single circuit such as a lighting circuit

(iv) reduce the possibility of unwanted tripping of RCDs due to excessive protective conductor currents produced by equipment in normal operation

(v) mitigate the effects of electromagnetic interferences (EMI)

(vi) prevent the indirect energizing of a circuit intended to be isolated.

11 Which one of the following is <u>not</u> a potential characteristic of equipment to be considered for harmful effects on other equipment?

- ○ a   Transient overvoltages
- ○ b   d.c. feedback
- ◉ c   Prospective short-circuit current
- ○ d   Excessive protective conductor current

Answer c

Refer to Regulation 331.1. Options a, b and d may be characteristics of equipment that might have harmful effects on other equipment, while option c is a characteristic of a point in the installation.

12 When making an assessment of the frequency and quality of maintenance, a factor to be considered is that

○ a   power factor is monitored
◉ b   protective measures for safety remain effective
○ c   starting currents are at a minimum
○ d   unbalanced loads need to be checked more frequently.

Answer b
Maintenance of the installation must be provided with sufficient frequency and to an appropriate standard so that it remains safe and suitable for continued use. See Chapter 34 (Maintainability), Regulation 341.1.

341.1 An assessment shall be made of the frequency and quality of maintenance the installation can reasonably be expected to receive during its intended life. The person or body responsible for the operation and/or maintenance of the installation shall be consulted. Those characteristics are to be taken into account in applying the requirements of Parts 4 to 7 so that, having regard to the frequency and quality of maintenance expected:

(i)   any periodic inspection and testing, maintenance and repairs likely to be necessary during the intended life can be readily and safely carried out, and

(ii)   **the effectiveness of the protective measures for safety during the intended life shall not diminish, and**

(iii)   the reliability of equipment for proper functioning of the installation is appropriate to the intended life.

Section 4

13 A BS 88-6 32 A fuse with U_0 of 230 V has a maximum tabulated Z_s for 5 seconds disconnection of

○ a   0.36 Ω
○ b   1.04 Ω
○ c   1.44 Ω
◉ d   1.84 Ω.

Answer d
You need to look at Table 41.4, which gives the maximum earth fault loop impedances for this nominal line voltage for this type of fuse.

Table 41.4 Maximum earth fault loop impedance (Z_S) for fuses, for 5 s disconnection time with U_0 of 230 V   (see Regulation 411.4.8)

(a) General purpose (gG) fuses to BS 88-2.2 and BS 88-6

Rating (amperes)	6	10	16	20	25	32	40	50
Z_S (ohms)	13.5	7.42	4.18	2.91	2.30	**1.84**	1.35	1.04

From the *IEE Wiring Regulations*, extract from Table 41.4, page 49

14 Which of the following is a requirement in terms of fault protection where the protective measure of electrical separation is employed?

- ⊙ a The voltage of the separated circuit is not more than 500 V
- ○ b Flexible cables and cords liable to mechanical damage are concealed
- ○ c Live parts are connected to a protective conductor
- ○ d Exposed-conductive-parts are connected to Earth

Answer a
Regulation 413.3.2 states that the voltage shall not exceed 500 V. Options b, c and d contravene the requirements of Regulations 413.3.3, 413.3.4 and 413.3.6.

413.3.2 The separated circuit shall be supplied through a source with at least simple separation, and the voltage of the separated circuit shall not exceed 500 V.

15 An RCD can be used as additional protection in the event of failure of the provision for basic protection when the RCD has a rating not exceeding

- ⊙ a 30 mA
- ○ b 100 mA
- ○ c 300 mA
- ○ d 500 mA.

Answer a
See Regulation 415.1.1, which shows that option a is the answer.

415.1.1 The use of RCDs with a rated residual operating current ($I_{\Delta n}$) not exceeding 30 mA and an operating time not exceeding 40 ms at a residual current of 5 $I_{\Delta n}$ is recognised in a.c. systems as additional protection in the event of failure of the provision for basic protection and/or the provision for fault protection or carelessness by users.

Notes

16 Basic protection can be provided by

- ○ a   protective earthing
- ◉ b   barriers
- ○ c   automatic disconnection in case of a fault
- ○ d   functional earth.

Answer b

Basic protection is the term given to those protective measures which prevent persons or livestock from coming into direct contact with live parts. Barriers can provide basic protection. See Regulations 416.2.1 to 416.2.5. Options a and c are components of fault protection and option d, 'functional earth', is not associated with safety.

17 An obstacle is not intended to

- ○ a   prevent unintentional bodily approach to live parts
- ◉ b   prevent intentional contact with live parts during operation
- ○ c   be removable by use of a key or tool
- ○ d   be secured so as to prevent unintentional removal.

Answer b

This is the subject of basic protection by the rule of obstacles. See Regulation 417.2 which shows that option b is correct.

417.2 Obstacles

NOTE: Obstacles are intended to prevent unintentional contact with live parts but not intentional contact by deliberate circumvention of the obstacle.

417.2.1 Obstacles shall prevent:
(i)   unintentional bodily approach to live parts, and
(ii)   unintentional contact with live parts during the operation of live equipment in normal service.

417.2.2 An obstacle may be removed without the use of a key or tool but shall be secured so as to prevent unintentional removal.

18 For a non-conducting location, the resistance of an insulating floor or wall at every point of measurement specified in Part 6 shall not be less than

- ○ a   45 kΩ where the voltage to earth does not exceed 230 V
- ◉ b   50 kΩ where the voltage to earth does not exceed 500 V
- ○ c   55 kΩ where the voltage to earth does not exceed 750 V
- ○ d   60 kΩ where the voltage to earth does not exceed 1000 V.

Answer b

Limits are set for the resistance of insulating floors or walls by Regulation 418.1.5. Protection by a non-conducting location is not recognised for general use: see Regulation 418.1.

418.1.5 The resistance of insulating floors and walls at every point of measurement under the conditions specified in Part 6 shall be not less than:

(i) **50 kΩ, where the nominal voltage of the installation does not exceed 500 V**, or

(ii) 100 kΩ, where the nominal voltage of the installation exceeds 500 V.

NOTE: If at any point the resistance is less than the specified value, the floors and walls are deemed to be extraneous-conductive-parts for the purposes of protection against electric shock.

19 Where electrical separation is used to supply more than one Item of current-using equipment

○ a   warning notices may be omitted

◉ b   the length of the wiring system should not exceed 500 m

○ c   socket-outlets should not have a protective conductor contact

○ d   the provisions for basic protection are not required.

Answer b

See Regulation 418.3.8, which states that the length of the wiring system should not exceed 500 m. Also see Regulations 418.3, 418.3.1 and 418.3.5 which show that options a, c and d are incorrect.

20 In locations with risks of fire due to the nature of processed or stored materials, enclosures such as heaters and resistors shall not attain in normal conditions higher surface temperatures than

◉ a   90 °C

○ b   85 °C

○ c   60 °C

○ d   55 °C.

Answer a

In normal locations, BS 7671 accepts the temperature limits set by the British Standard for the particular item of equipment. However, in locations with risks of fire due to the nature of processed or stored materials, BS 7671 sets limits that override equipment standards. See Regulation 422.3.2.

Notes

422.3.2 Measures shall be taken to prevent an enclosure of electrical equipment such as a heater or resistor from exceeding the following temperatures:

(i) **90 °C under normal conditions**, and

(ii) 115 °C under fault conditions.

21 In a paper mill installation wired in PVC single-core cables in conduit and trunking, where a fire may be caused by a resistive fault in overhead heaters with film elements, protection against an insulation fault shall be provided by a

- ○ a   BS 88 fuse
- ○ b   BS EN 60898 cb
- ○ c   300 mA RCD
- ◉ d   30 mA RCD.

Answer d

See Regulation 422.3.9, second paragraph of (i).

422.3.9 Except for mineral insulated cables, busbar trunking systems or powertrack systems, a wiring system shall be protected against insulation faults:

(i)   in a TN or a TT system, by an RCD having a rated residual operating current ($I_{\Delta n}$) not exceeding 300 mA according to Regulation 531.2.4 and to relevant product standards.
Where a resistive fault may cause a fire, eg for overhead heating with heating film elements, the rated residual operating current shall not exceed 30 mA.

(ii)   in an IT system, by an insulation monitoring device with audible and visual signals provided in accordance with Regulation 538.1. Disconnection times in the event of a second fault are given in Chapter 41.

22 The maximum temperature for the exposed metallic casing of a wall-mounted, cord-operated infra-red heater is

- ○ a   55 °C
- ○ b   65 °C
- ○ c   70 °C
- ◉ d   80 °C.

Answer d

See Section 423, Protection against burns, Table 42.1.

Table 42.1 Temperature limit under normal load conditions for an accessible part of equipment within arm's reach

Accessible part	Material of accessible surfaces	Maximum temperature (°C)
A hand-held part	Metallic	55
	Non-metallic	65
A part intended to be touched but not hand-held	Metallic	70
	Non-metallic	80
A part which need not be touched for normal operation	**Metallic**	**80**
	Non-metallic	90

From the *IEE Wiring Regulations*, Table 42.1, page 69

23 Where fault current protection is <u>not</u> required, protection against overload current is generally provided by

○ a   BS EN 60898 fuses
◉ b   an inverse-time-lag protective device
○ c   a BS 3036 semi-enclosed fuse
○ d   RCDs.

Answer b

Options a and c will provide both overload and fault protection and option d provides protection against earth faults. See Regulation 432.2.

432.2 Protection against overload current only

A device providing protection against overload current is generally an inverse-time-lag protective device whose rated short-circuit breaking capacity may be below the value of the maximum prospective fault current at the point where the device is installed. Such a device shall satisfy the relevant requirements of Section 433.

24 Regulations 434.2 and 434.2.1 do <u>not</u> allow the fault current protective device to be placed on the load side of the reduction in current-carrying capacity of a conductor if

○ a   the length of conductor is less than 3 m

◉ b   the length of conductor exceeds 3 m

○ c   it is erected in such a manner as to reduce to a minimum the risk of fault current

○ d   it is erected in such a manner as to reduce to a minimum the risk of fire or danger to persons.

Answer b

BS 7671 allows the fault current protective device to be placed on the load side of the reduction in current-carrying capacity of a conductor as is often necessary in short cable lengths between busbars and the overcurrent device, providing precautions are taken, such as ensuring that the cable length is no more than 3 metres. See Section 434 and Regulations 434.2 and 434.2.1.

434.2 Position of devices for protection against fault current

Except where Regulation 434.2.1, 434.2.2 or 434.3 applies, a device providing protection against fault current shall be installed at the point where a reduction in the cross-sectional area or other change causes a reduction in the current-carrying capacity of the conductors of the installation.

434.2.1 The regulatons in Regulation 434.2 shall not be applied to installations situated in locations presenting a fire risk or risk of explosion and where the requirements for special installations and locations specify different conditions.

Except where Regulation 434.2.2 or 434.3 applies, a device for protection against fault current may be installed other than as specified in Regulation 434.2, under the following conditions:

The part of the conductor between the point of reduction of cross-sectional area or other change and the position of the protective device shall:
(i)   not exceed 3 m in length, and
(ii)   be installed in such a manner as to reduce the risk of fault to a minimum, and
     NOTE: This condition may be obtained, for example, by reinforcing the protection of the wiring against external influences.
(iii) be installed in such a manner as to reduce to a minimum the risk of fire or danger to persons.

25 In the formula

$$t = \frac{k^2 S^2}{I^2}$$

what is the k factor for copper conductors insulated with 70 °C thermoplastic and an assumed initial temperature of 70 °C?

- ○ a   100/86
- ◉ b   115/103
- ○ c   141
- ○ d   143

Answer b
Refer to Regulation 434.5.2 and Table 43.1. Be careful that you read off the value for 70 °C thermoplastic.

Table 43.1   Values of k for common materials, for calculation of the effects of fault current for disconnection times up to 5 seconds

Conductor insulation	Thermoplastic				Thermosetting		Mineral insulated	
	90 °C		70 °C		90 °C	60 °C	Thermoplastic sheath	Bare (unsheathed)
Conductor cross-sectional area	≤ 300 mm²	> 300 mm²	≤ 300 mm²	> 300 mm²				
Initial temperature	90 °C		70 °C		90 °C	60 °C	70 °C	105 °C
Final temperature	160 °C	140 °C	160 °C	140 °C	250 °C	200 °C	160 °C	250 °C
Copper conductor	k = 100	k = 86	**k = 115**	**k = 103**	k = 143	k = 141	k = 115	k = 135/115[a]
Aluminium conductor	k = 66	k = 57	k = 76	k = 68	k = 94	k = 93		
Tin soldered joints in copper conductors	k = 115	k = 115	k = 115	k = 115	k = 115	k = 115		

[a] This value shall be used for bare cables exposed to touch.

From the *IEE Wiring Regulations*, Table 43.1, page 77

Notes

26 **What is the required minimum impulse withstand voltage for category IV equipment for use in an installation with nominal voltage of 400/690?**

- ○ a   2.5 kV
- ○ b   4 kV
- ○ c   6 kV
- ◉ d   8 kV

Answer d

You need to refer to Section 443 and Table 44.3 for minimum impulse withstand voltage.

Table 44.3   Required minimum impulse withstand voltage

Nominal voltage of the installation V	Required minimum impulse withstand voltage kV [1]			
	Category IV (equipment with very high impulse voltage)	Category III (equipment with high impulse voltage)	Category II (equipment with normal impulse voltage)	Category I (equipment with reduced impulse voltage)
230/240 277/480	6	4	2.5	1.5
400/690	**8**	6	4	2.5
1000	12	8	6	4

From the *IEE Wiring Regulations*, Table 44.3, page 85

[1] This impulse withstand voltage is applied between live conductors and PE.

Section 5

27 **When selecting equipment for an installation, the person who makes the final decision over its safety would be the**

- ○ a   representative of a regional supply company
- ○ b   representative of a national inspection body
- ◉ c   designer or other person responsible
- ○ d   inspector and tester.

Answer c

Refer to Section 511 (Compliance with Standards). If equipment to be installed is 'not covered' by a British Standard and there is no standard for the particular item of equipment, then Regulation 511.2 provides guidance. The designer is ultimately responsible for selection.

511.2 Where equipment to be used is not covered by a British Standard or Harmonized Standard or is used outside the scope of its standard, the designer or other person responsible for specifying the installation shall confirm that the equipment provides the same degree of safety as that afforded by compliance with the Regulations.

28 Unless otherwise confirmed as suitable, switchgear, protective devices and accessories shall not be connected to conductors intended to operate at a temperature in excess of

- ○ a   30 °C
- ○ b   50 °C
- ◉ c   70 °C
- ○ d   90 °C.

Answer c

While thermosetting cables with a conductor operating temperature of 90 °C are commonly available, most equipment is only designed for cables operating at a maximum of 70 °C (option c). Before using a cable loaded so that its conductor temperature exceeds 70 °C, it must be checked with the supplier that the switchgear etc is suitable for the temperature. It may be necessary to increase the cable size to reduce the conductor temperature. Refer to Regulation 512.1.2.

512.1.2 Current

Every item of equipment shall be suitable for:
- (i)   the design current, taking into account any capacitive and inductive effects, and
- (ii) the current likely to flow in abnormal conditions for such periods of time as are determined by the characteristics of the protective devices concerned.

Switchgear, protective devices, accessories and other types of equipment shall not be connected to conductors intended to operate at a temperature exceeding 70 °C at the equipment in normal service, unless the equipment manufacturer has confirmed that the equipment is suitable for such conditions.

Notes

29 A 70 °C thermoplastic insulated and sheathed cable to BS 6004 is installed in an ambient temperature of 25 °C for part of the run and then enters an area at 40 °C. The effect on the cable will be to

- ○ a   require clips at greater intervals
- ○ b   reduce the voltage drop
- ○ c   increase the bending radius
- ◉ d   decrease its current-carrying capacity.

Answer d

Refer to Table 4B1. This shows that increasing the ambient temperature makes more severe the ambient rating factor and, consequently, reduces the current-carrying capacity. The current-carrying capacity of a cable is limited by the maximum operating temperature of the cable. The more current that flows in a cable, the hotter it gets as a result of the I^2R heating effect of the current.

Each current rating table in Appendix 4 is headed by the ambient temperature at which the rating was determined and the maximum conductor operating temperature used to determine the rating.

From the *IEE Wiring Regulations*, extract from Table 4B1, page 267

Table 4B1   Rating factor for ambient air temperatures other than 30 °C to be applied to the current-carrying capacities for cables in free air

Ambient temperature °C	Insulation			
	70 °C thermoplastic	90 °C thermosetting	Mineral	
			Thermoplastic covered or bare and exposed to touch 70 °C	Bare and not exposed to touch 105 °C
25	1.03	1.02	1.07	1.04
30	1.00	1.00	1.00	1.00
35	0.94	0.96	0.93	0.96
40	0.87	0.91	0.85	0.92
45	0.79	0.87	0.78	0.88
50	0.71	0.82	0.67	0.84
55	0.61	0.76	0.57	0.80
60	0.50	0.71	0.45	0.75

30 A single-phase load of 13 A is supplied via a 15 A BS 1361 fuse using single-core 70 °C thermoplastic copper cables installed to method 4 (Reference Method B). The rating factor for grouping is 0.7 and for ambient temperature 0.87, and overload protection is to be provided. The minimum acceptable size cable would be

Notes

- ○ a 1 mm²
- ○ b 1.5 mm²
- ○ c 2.5 mm²
- ◉ d 4 mm².

Answer d

Refer to Appendix 4, 5.1.2 for the calculation.

The procedure is:
(1) the design current I_b of the circuit must be established (I_b is 13 A)
(2) the overcurrent device rating I_n is then selected so that I_n is greater than or equal to I_b

From the question, the device rating I_n is 15 A, which is greater than the load I_b which is 13 A, so this is satisfactory.

The required tabulated current-carrying capacity of the cable is given by $I_t \geq \dfrac{I_n}{C_a C_i C_g C_c}$

where:
C_a is the rating factor for ambient temperature
C_i is the rating factor for thermal insulation
C_g is the rating factor for grouping
C_c is the rating factor for semi-enclosed fuses

In this example C_a is 0.87, and C_g is 0.7. C_i can be assumed to be 1 as no reference to thermal insulation is made and C_c is 1 as rewirable fuses are not used.

Hence $I_t = \dfrac{15}{0.87 \times 0.7} = 24.63$ A

From Table 4D1A of BS 7671, column 4, the cable ratings are:

1 mm²	13.5 A
1.5 mm²	17.5 A
2.5 mm²	24 A
4 mm²	32 A

As 2.5 mm² has slightly too low a rating when overcurrent protection is to be provided, the answer must be option d.

Notes

31 The minimum information contained on a periodic inspection and testing notice would be the

- ○ a   date of inspection and inspector's name
- ○ b   date of next inspection and client's name
- ◉ c   date of last inspection and date of next inspection
- ○ d   company's address and date of next inspection.

Answer c

The information required is stated in Regulation 514.12.1.

514.12.1   A notice of such durable material as to be likely to remain easily legible throughout the life of the installation, shall be fixed in a prominent position at or near the origin of every installation upon completion of the work carried out in accordance with Chapter 61 or 62. The notice shall be inscribed in indelible characters not smaller than those illustrated here and shall read as follows:

IMPORTANT

This installation should be periodically inspected and tested and a report on its condition obtained, as prescribed in the IEE Wiring Regulations BS 7671 Requirements for Electrical Installations.

Date of last inspection ....

Recommended date of next inspection ....

---

32 When conductors are identified by numbers, the neutral shall be identified by the number

- ◉ a   0
- ○ b   4
- ○ c   5
- ○ d   14.

Answer a

You need to refer to Regulation 514.5.4.

514.5.4 Numeric
Conductors may be identified by numbers, the number 0 being reserved for the neutral or midpoint conductor.

33 Which one of the following devices would <u>not</u> be suitable to provide overcurrent protection?

- ⊙ a   Residual current device to BS 4293
- ○ b   Cartridge fuse to BS 1362
- ○ c   Cartridge fuse to BS 88-6
- ○ d   Rewirable fuse to BS 3036

Answer a
Refer to Part 2 Definitions, which gives the definitions of fuses and residual current devices. Fuses provide overcurrent protection, as the fuse element melts to limit the current. RCDs do not provide overcurrent protection. A residual current device only operates when the algebraic sum of the neutral and line currents is not zero; it will not react to an overcurrent.

Fuse. A device which, by the melting of one or more of its specially designed and proportioned components, opens the circuit in which it is inserted by breaking the current when this exceeds a given value for a sufficient time. The fuse comprises all the parts that form the complete device.

Residual current device (RCD). A mechanical switching device or association of devices intended to cause the opening of the contacts when the residual current attains a given value under specified conditions.

34 Under fault conditions, cables are subjected to stress and thermal effects. One item <u>not</u> normally encountered would be

- ⊙ a   electrochemical effects
- ○ b   electromechanical stresses
- ○ c   electromagnetic effects
- ○ d   thermal damage.

Answer a
BS 7671 contains specific requirements with regard to electromechanical stresses, electromagnetic effects and thermal damage (options b, c and d) likely to occur under fault conditions. Option a, electrochemical effects, such as electrolytic corrosion, are only mentioned in relation to the selection and erection of wiring systems and earth arrangements. See Regulations 521.5.1 and 521.5.2.

Notes

35 Every connection between conductors, or between conductors and equipment, need <u>not</u> be selected to account for the

- ⊙ a   conductor length
- ○ b   material of the conductor
- ○ c   number and shape of the wires forming the conductor
- ○ d   cross-sectional area of the conductor.

Answer a

In Section 526, Regulation 526.2 lists a number of considerations, of which options b, c and d are three. Option a is not mentioned as it has no bearing on the suitability of a connection for use.

526.2   The selection of the means of connection shall take account of, as appropriate:
(i)   the material of the conductor and its insulation
(ii)   the number and shape of the wires forming the conductor
(iii)   the cross-sectional area of the conductor
(iv)   the number of conductors to be connected together
(v)   the temperature attained at the terminals in normal service such that the effectiveness of the insulation of the conductors connected to them is not impaired
(vi)   the provision of adequate locking arrangements in situations subject to vibration or thermal cycling.

Where a soldered connection is used the design shall take account of creep, mechanical stress and temperature rise under fault conditions.

36 Where more than one firefighter's switch is installed on any one building, each switch must be

- ○ a   not more than 3.75 m from the ground
- ○ b   in a locked location to prevent nuisance operation
- ○ c   electrically linked
- ⊙ d   clearly marked.

Answer d

The requirements for firefighter's switches are in Regulation group 537.6. Where there is more than one, each switch must be clearly marked to indicate what it will switch. See Regulation 537.6.3, item (iv).

537.6.3 Every firefighter's switch provided for compliance with Regulation 537.6.1 shall comply with all the relevant requirements of the following items (i) to (iv) and any requirements of the local fire authority:

(i) For an exterior installation, the switch shall be outside the building and adjacent to the equipment, or alternatively a notice indicating the position of the switch shall be placed adjacent to the equipment and a notice shall be fixed near the switch so as to render it clearly distinguishable

(ii) For an interior installation, the switch shall be in the main entrance to the building or in another position to be agreed with the local fire authority

(iii) The switch shall be placed in a conspicuous position, reasonably accessible to firefighters and, except where otherwise agreed with the local fire authority, at not more than 2.75 m from the ground or the standing beneath the switch

(iv) Where more than one switch is installed on any one building, each switch shall be clearly marked to indicate the installation or part of the installation which it controls.

37 Where safety depends upon the direction of rotation of a motor, provision shall be made to

○ a   measure the maximum and minimum speed
○ b   monitor the slip frequency
◉ c   prevent reverse operation
○ d   prevent operation of overload devices.

Answer c

Looking up 'Motors – direction of rotation' in the index directs you to Regulation 537.5.4.3, which defines option c as the answer.

537.5.4.3 Where safety depends on the direction of rotation of a motor, provision shall be made for the prevention of reverse operation due to, for example, a reversal of phases.

Notes

38 A 16 mm² multicore cable incorporating a copper protective conductor is insulated with 90 °C thermosetting insulation. Given a fault current of 500 A and a disconnection time of 0.4 second, the minimum acceptable size of cpc would be

○ a   1.5 mm²
◉ b   2.5 mm²
○ c   4.0 mm²
○ d   6.0 mm².

Answer b

See Regulation 543.1.3. The fault current I is given in the question, as is the disconnection time t. k must be looked up in Table 54.3. From Table 54.3, the k value for a copper protective conductor incorporated in a cable with 90 °C thermosetting insulation is 143.

Hence:

$$S = \frac{\sqrt{500^2 \times 0.4}}{143} = 2.211 \text{ mm}^2$$

The last sentence of Regulation 543.1.3 states: Where the application of the formula produces a non-standard size, a conductor having the nearest larger standard cross-sectional area shall be used.

This is 2.5 mm².

543.1.3 The cross-sectional area, where calculated, shall be not less than the value determined by the following formula or shall be obtained by reference to BS 7454:

$$S = \frac{\sqrt{I^2 t}}{k}$$

Table 54.3 Values of k for protective conductor incorporated in a cable or bunched with cables, where the assumed initial temperature is 70 °C or greater

Notes

Material of conductor	Insulation material		
	70 °C thermoplastic	90 °C thermoplastic	90 °C thermosetting
Copper	115/103*	100/86*	**143**
Aluminium	76/68*	66/57*	94
Assumed initial temperature	70 °C	90 °C	90 °C
Final temperature	160 °C/140 °C*	160 °C/140 °C*	250 °C

* Above 300 mm^2

From the *IEE Wiring Regulations*, Table 54.3, page 129

39 When using the alternative method to Regulation 543.1.3 of sizing a copper protective conductor, the minimum size to be used when a copper line conductor has a csa of 50 mm^2 would be

○ a   16 mm^2
◉ b   25 mm^2
○ c   35 mm^2
○ d   50 mm^2.

Answer b

See Regulation group 543.1 (Cross-sectional areas) for advice on how to determine protective conductor sizes. Regulation 543.1.1 offers two approaches. The first method is calculation using the formula. The alternative method is selection in accordance with Regulation 543.1.4 and Table 54.7.

In the question, S (at 50 mm^2) > 35, so from Table 54.7 the minimum protective conductor size is S/2, that is 50/2 or 25mm^2, therefore option b is the answer.

Notes

543.1.1 The cross-sectional area of every protective conductor, other than a protective bonding conductor, shall be:
(i)   calculated in accordance with Regulation 543.1.3, or
(ii)   selected in accordance with Regulation 543.1.4.

Calculation in accordance with Regulation 543.1.3 is necessary if the choice of cross-sectional area of line conductors has been determined by considerations of short-circuit current and if the earth fault current is expected to be less than the short-circuit current.

543.1.4 Where it is desired not to calculate the minimum cross-sectional area of a protective conductor in accordance with Regulation 543.1.3, the cross-sectional area may be determined in accordance with Table 54.7.

Where the application of Table 54.7 produces a non-standard size, a conductor having the nearest larger standard cross-sectional area shall be used.

Table 54.7   Minimum cross-sectional area of protective conductor in relation to the cross-sectional area of associated line conductor

Cross-sectional area of line conductor S	Minimum cross-sectional area of the corresponding protective conductor	
	If the protective conductor is of the same material as the line conductor	If the protective conductor is not of the same material as the line conductor
(mm²)	(mm²)	(mm²)
$S \leq 16$	S	$\dfrac{k1}{k2} \times S$
$16 < S \leq 35$	16	$\dfrac{k1}{k2} \times 16$
$S > 35$	$\dfrac{S}{2}$	$\dfrac{k1}{k2} \times \dfrac{S}{2}$

From the *IEE Wiring Regulations*, Table 54.7, page 130

40 The maximum current rating of a protective device protecting a circuit including B22 bayonet cap lampholders would be

- ○ a 5 A
- ○ b 6 A
- ● c 16 A
- ○ d 20 A.

Answer c

On the failure of a lamp, a short circuit can develop, which may cause an electric arc that needs to be interrupted by the overcurrent device (before significant overheating of the luminaire occurs). Consequently, a maximum device rating is set by Regulation 559.6.1.6.

559.6.1.6 Lighting circuits incorporating B15, B22, E14, E27 or E40 lampholders shall be protected by an overcurrent protective device of maximum rating 16 A.

41 Which of the following is <u>not</u> a BS 7671 requirement when installing a heating cable laid directly in soil?

- ○ a It is completely embedded in the soil
- ○ b It does not suffer damage in the event of normal movement
- ○ c It complies with the manufacturer's instructions
- ● d It is protected by a 500 mA RCD

Answer d

This is typical of under-soil heating in a horticultural location for soil warming for plant and seed propagation. See Regulation 554.4.3, where options a, b and c are all stated as requirements.

554.4.3 A heating cable laid directly in soil, a road or the structure of a building shall be installed so that it:
(i)   is completely embedded in the substance it is intended to heat, and
(ii)   does not suffer damage in the event of movement normally to be expected in it or the substance in which it is embedded, and
(iii) complies in all respects with the manufacturer's instructions and recommendations.

Notes

Section 6

42 During the installation inspection process, which one of the following need not be checked?

- ⦿ a   The external earth fault loop impedance
- ○ b   Equipment in compliance with Section 511
- ○ c   Correct selection and erection
- ○ d   Visible damage or defects to installed equipment

Answer a
Earth fault loop impedance is an instrument test, and is not part of the inspection process.

43 When testing SELV circuits, the minimum acceptable insulation resistance value is

- ○ a   0.25 MΩ
- ⦿ b   0.5 MΩ
- ○ c   1 MΩ
- ○ d   2 MΩ.

Answer b
You can find the requirements for minimum values of insulation resistance in Table 61.

612.3.2   The insulation resistance measured with the test voltages indicated in Table 61 shall be considered satisfactory if the main switchboard and each distribution circuit tested separately, with all its final circuits connected but with current-using equipment disconnected, has an insulation resistance not less than the appropriate value given in Table 61.

NOTE: More stringent requirements are applicable for the wiring of fire alarm systems in buildings, see BS 5839-1.

Table 61   Minimum values of insulation resistance

Circuit nominal voltage (V)	Test voltage d.c. (V)	Minimum insulation resistance (MΩ)
SELV and PELV	250	**≥ 0.5**
Up to and including 500 V with the exception of the above systems	500	≥ 1.0
Above 500 V	1000	≥ 1.0

From the *IEE Wiring Regulations*, Table 61, page 158

44 Defects identified during an initial verification should be

- ○ a   noted on the certificate, and the client should be informed
- ○ b   pointed out to the main contractor when the certificate has been completed
- ◉ c   made good before the certificate is issued
- ○ d   put on a snagging list to be rectified later.

Answer c

The general requirements for initial testing are given in Regulation 612.1. See the fifth paragraph. Note that the fault must be rectified and the test must be repeated as well as any others that may have been influenced by the fault.

612.1 General

... If any test indicates a failure to comply, that test and any preceding test, the results of which may have been influenced by the fault indicated, shall be repeated after the fault has been rectified ...

45 Which of the following is <u>not</u> a requirement for periodic inspection and testing?

- ○ a   Ensuring the safety of persons and livestock from shock and burns
- ○ b   Ensuring the property is protected against damage from fire and heat due to an installation fault
- ○ c   Confirmation that the installation is not damaged as to impair safety
- ◉ d   Confirmation that the supply to the installation meets all the requirements of the Electricity Safety, Quality and Continuity Regulations

Answer d

Options a, b and c are listed in Regulation 621.2. Option d is not a requirement. It is the responsibility of the Distribution Network Operator (DNO) to confirm full compliance of the supply.

621.2 Periodic inspection comprising a detailed examination of the installation shall be carried out without dismantling, or with partial dismantling as required, supplemented by appropriate tests from Chapter 61 to show that the requirements for disconnection times, as set out in Chapter 41 for protective devices, are complied with, to provide for:
- (i)   safety of persons and livestock against the effects of electric shock and burns
- (ii)   protection against damage to property by fire and heat arising from an installation defect

Notes

(iii)   confirmation that the installation is not damaged or deteriorated so as to impair safety

(iv)   the identification of installation defects and departures from the requirements of these Regulations that may give rise to danger.

46 Additional work is carried out to an installation, which comprises a socket-outlet added to an existing circuit. The paperwork to be completed would be

○ a   a Periodic Inspection Report
○ b   a Schedule of Test Results
◉ c   a Minor Electrical Installation Works Certificate
○ d   a Schedule of Inspections.

Answer c
You will find that Chapter 63(Certification and Reporting), Section 631 (General) gives the requirements for the issue of certificates. Regulation 631.3 gives guidance on minor works (option c). It is to be noted that a Minor Electrical Installation Works Certificate is to be issued for each (and every) circuit altered or extended. There is also guidance in Appendix 6 in the notes to the Minor Electrical Installation Works Certificate.

631.3   Where minor electrical installation work does not include the provision of a new circuit, a Minor Electrical Installation Works Certificate, based on the model given in Appendix 6, may be provided for each circuit altered or extended as an alternative to an Electrical Installation Certificate.

Appendix 6 Minor Electrical Installation Works Certificate
NOTES:
The Minor Works Certificate is intended to be used for additions and alterations to an installation that do not extend to the provision of a new circuit. Examples include the addition of socket-outlets or lighting points to an existing circuit, the relocation of a light switch etc. This Certificate may also be used for the replacement of equipment such as accessories or luminaires, but not for the replacement of distribution boards or similar items. Appropriate inspection and testing, however, should always be carried out irrespective of the extent of the work undertaken.

47 When completing an Electrical Installation Certificate, the person who does <u>not</u> have to sign the certificate would be the

○ a   tester
◉ b   client
○ c   constructor
○ d   designer.

Answer b

See the advice is given in the introduction to the model forms (Appendix 6). There is no need for the client to sign as they will receive a copy of the certificate. See also Regulation 632.3.

Section 7

48 For an electric floor heating system in a bathroom, a fine mesh metallic grid is <u>not</u> required to be connected to the protective conductor of the supply circuit if the protective measure used is

◉ a   SELV
○ b   electrical separation
○ c   a non-conducting location
○ d   an earth-free equipotential zone.

Answer a

Option b is not permitted in these circumstances (see Regulation 701.753). Options c and d are only for use where the installation is controlled or under the supervision of skilled or instructed persons, and so would not be permitted in a bathroom (see Regulations 701.410.3.6, 418.1 and 418.2).

701.753 Electric floor heating systems

For electric floor heating systems, only heating cables according to relevant product standards or thin sheet flexible heating elements according to the relevant equipment standard shall be erected provided that they have either a metal sheath or a metal enclosure or a fine mesh metallic grid. The fine mesh metallic grid, metal sheath or metal enclosure shall be connected to the protective conductor of the supply circuit. Compliance with the latter requirement is not required if the protective measure SELV is provided for the floor heating system.

For electric floor heating systems the protective measure "protection by electrical separation" is not permitted.

49 Access to the space under a bath is gained by unscrewing the bath panel retaining clips. This space is classified as

- ○ a   zone 0
- ○ b   zone 1
- ◉ c   outside the zones
- ○ d   zone 2.

Answer c

A screwdriver (tool) is required for the panel removal, making the space outside the zones. See Regulation 701.32.3.

701.32.3 Description of zone 1

Zone 1 is limited by:

(i)   the finished floor level and the horizontal plane corresponding to the highest fixed shower head or water outlet or the horizontal plane lying 2.25 m above the finished floor level, whichever is higher

(ii) the vertical surface:
   a)   circumscribing the bath tub or shower basin (see Figure 701.1)
   b)   at a distance of 1.20 m from the centre point of the fixed water outlet on the wall or ceiling for showers without a basin (see Figure 701.2 (e) and (f)).

Zone 1 does not include zone 0.

The space under the bath tub or shower basin is considered to be zone 1. **However, if the space under the bath tub or shower basin is only accessible with a tool, it is considered to be outside the zones.**

50 For SELV and PELV circuits in a bathroom, basic protection for equipment in all zones shall be provided by

- ○ a   obstacles
- ◉ b   barriers or enclosures
- ○ c   non-conducting location
- ○ d   placing out of reach.

Answer b

See Regulation 701.414.4.5, item (ii), which shows that option b is correct. Options a, c and d are not permitted in a bathroom (see Regulations 701.410.3.5 and 701.410.3.6).

701.414.4.5 Requirements for SELV and PELV circuits

Where SELV or PELV is used, whatever the nominal voltage, basic protection for equipment in zones 0, 1 and 2 shall be provided by:

(i)   basic insulation complying with 416.1, or

(II)   barriers or enclosures complying with 416.2 and affording a degree of protection of at least IPXXB or IP2X.

---

51 Electrical equipment in zones 0 and 1 of a fountain shall have mechanical protection to a minimum classification of

○ a   AG0

○ b   AG1

◉ c   AG2

○ d   AG3.

Answer c

In Regulation 702.55.3 the impact severity is given as medium, ie AG2. Option a does not exist and options b and d are too low and too high respectively. See Appendix 5 for definitions of AG1–3.

702.55.3 Electrical equipment of fountains

Electrical equipment in zones 0 or 1 shall be provided with mechanical protection to medium severity (AG2), eg by use of mesh glass or by grids which can only be removed by the use of a tool.

A luminaire installed in zones 0 or 1 shall be fixed and shall comply with BS EN 60598-2-18.

An electric pump shall comply with the requirements of BS EN 60335-2-41.

---

52 Which of the following would not be a requirement for an Assembly for Construction Sites (ACS), supplying mobile current-using equipment on a construction site?

◉ a   Main equipotential bonding

○ b   Overcurrent protective devices

○ c.   Devices for fault protection

○ d   Socket outlets

Answer a

Options b, c and d are all requirements of Regulation 704.537.2.2. Option a is a process, not a device or accessory.

704.537.2.2 Devices for isolation

Each Assembly for Construction Sites (ACS) shall incorporate suitable devices for the switching and isolation of the incoming supply.

A device for isolating the incoming supply shall be suitable for securing in the off position (see Regulation 537.2.1.5), for example, by padlock or location of the device inside a lockable enclosure.

Current-using equipment shall by supplied by ACSs, each ACS comprising:
(i)   overcurrent protective devices, and
(ii)   devices affording fault protection, and
(iii)   socket-outlets, if required.
Safety and standby supplies shall be connected by means of devices arranged to prevent interconnection of the different supplies.

53 Which one of the following is not included in the scope of Section 705?

○ a   Chicken-houses
○ b   Stables
○ c   Piggeries
◉ d   Staff residences

Answer d

Section 705 covers 'Agricultural and Horticultural Premises' and the scope is given in the first regulation. Staff premises are excluded, where the standard requirements apply.

705.1 Scope

The particular requirements of this section apply to fixed electrical installations indoors and outdoors in agricultural and horticultural premises. Some of the requirements are also applicable to other locations that are in common buildings belonging to the agricultural and horticultural premises. Where the special requirements also apply to residences and other locations in such common buildings this is stated in the text of the relevant regulations.

Rooms, locations and areas for household applications and similar are not covered by this section.

54 In conducting locations with restricted movement, supplies to handlamps shall be provided by

- ◉ a   SELV
- ○ b   reduced low voltage supplies
- ○ c   RCD or RCBO
- ○ d   PELV.

Answer a

The requirements for handlamps in restrictive conductive locations are given in Regulation 706.410.3.10 , item (ii).

Reduced low voltage and RCDs (options b and c) are not allowed. PELV (option d) is only permitted for the supply to fixed equipment.

55 The minimum height of overhead conductors on caravan sites, when installed in vehicle movement areas, is

- ○ a   3.0 m
- ○ b   3.5 m
- ○ c   5.8 m
- ◉ d   6.0 m.

Answer d

The requirements for electrical installations in caravan parks are given in Section 708. Where overhead conductors are installed they must be at least 6 m high in vehicle movement areas; see Regulation 708.521.1.2.

708.521.1.2 Overhead distribution circuits

All overhead conductors shall be insulated.

Poles and other supports for overhead wiring shall be located or protected so that they are unlikely to be damaged by any foreseeable vehicle movement.

Overhead conductors shall be at a height above ground of not less than 6 m in all areas subject to vehicle movement and 3.5 m in all other areas.

Notes

56 The particular requirements for mobile or transportable units apply to

○ a   generating sets
◉ b   mobile workshops
○ c   mobile machinery to BS EN 60204-1
○ d   traction equipment of electric vehicles.

Answer b

See Regulation 717.1, which shows that options a, c and d are not applicable.

717.1 Scope

The particular requirements of this section are applicable to mobile or transportable units.

For the purposes of this section, the term "unit" is intended to mean a vehicle and/or mobile or transportable structure in which all or part of an electrical installation is contained, which is provided with a temporary supply by means of, for example, a plug and socket-outlet.

Units are either:
(i)   of the mobile type, eg vehicles (self-propelled or towed), or
(ii)   of the transportable type, eg containers or cabins.
Examples of the units include technical and facilities vehicles for the entertainment industry, medical services, advertising, fire fighting, **workshops**, offices, transportable catering units etc.

The requirements are not applicable to:
(iii)   generating sets
(iv)   marinas and pleasure craft
(v)   mobile machinery in accordance with BS EN 60204-1
(vi)   caravans to Section 721
(vii)   traction equipment of electric vehicles
(viii)   electrical equipment required by a vehicle to allow it to be driven safely or used on the highway.
Additional requirements shall be applied where necessary for units including showers, or for medical locations, etc.

57 Electrical equipment in a circus installation must have a degree of protection of at least

○ a   IP33
○ b   IP4X
◉ c   IP44
○ d   IPX8.

Answer c

Regulation 740.512.2 indicates that electrical equipment in a circus must have a degree of protection of at least IP44. This is similar to that for agricultural locations where there is also livestock.

Section 8

58 The ambient air temperature rating factor for 90 °C thermosetting cables operating in an ambient air temperature of 60 °C is

- ○ a   0.50
- ○ b   0.56
- ○ c   0.63
- ◉ d   0.71.

Answer d

Correction factors for ambient air temperature are given in Table 4B1 of Appendix 4.

Table 4B1   Rating factors for ambient air temperatures other than 30 °C to be applied to the current-carrying capacities for cables in free air

| Ambient temperature °C | Insulation | | | |
| | 70 °C thermoplastic | 90 °C thermosetting | Mineral | |
			Thermoplastic covered or bare and exposed to touch 70 °C	Bare and not exposed to touch 105 °C
25	1.03	1.02	1.07	1.04
30	1.00	1.00	1.00	1.00
35	0.94	0.96	0.93	0.96
40	0.87	0.91	0.85	0.92
45	0.79	0.87	0.78	0.88
50	0.71	0.82	0.67	0.84
55	0.61	0.76	0.57	0.80
60	0.50	**0.71**	0.45	0.75

From the *IEE Wiring Regulations*, extract from Table 4B1, page 267

Notes

59 A BS 3036 100 A fuse, when carrying 1800 A, has a disconnection time of

- ⦿ a   0.2 second
- ○ b   0.4 second
- ○ c   4 seconds
- ○ d   5 seconds.

Answer a

Overcurrent device characteristics are given in Appendix 3. Figures 3.2A and B give data for BS 3036 fuses (see Figure 3.2B, opposite).

60 A 2.5 mm² thermoplastic insulated and sheathed flat cable with protective conductor is laid in a ceiling beneath thermal insulation 80 mm thick in contact with the ceiling board, as shown in the figure below. What is its installed rating?

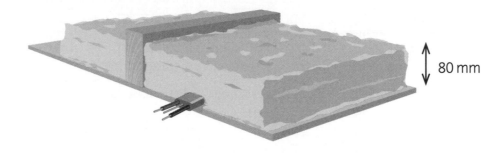

80 mm

- ○ a   17 A
- ○ b   20 A
- ⦿ c   21 A
- ○ d   27 A

Answer c

Refer to Table 4A2 of Appendix 4. Reference method 100 is appropriate in this instance.

Refer then to Table 4D5, which gives the cable rating for reference method 100 and 2.5 mm² cable as 21A.

Notes

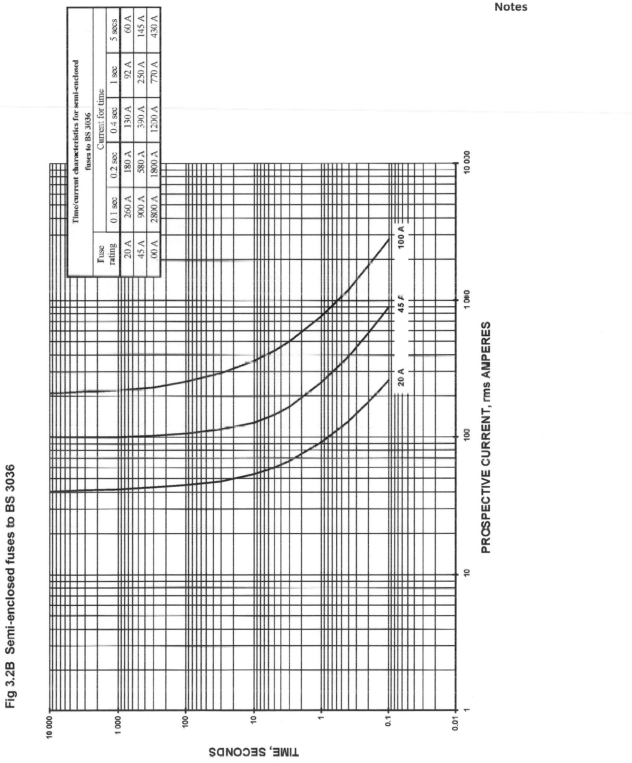

Fig 3.2B Semi-enclosed fuses to BS 3036

Time/current characteristics for semi-enclosed fuses to BS 3036					
Fuse rating	Current for time				
	0.1 sec	0.2 sec	0.4 sec	1 sec	5 secs
20 A	260 A	180 A	130 A	92 A	60 A
45 A	900 A	580 A	390 A	250 A	145 A
00 A	2800 A	1800 A	1200 A	770 A	430 A

From the *IEE Wiring Regulations*,
Figure 3.2B, page 246

Notes

Answer key

Sample test 2

Question	Answer	Question	Answer
1	a	31	c
2	c	32	a
3	b	33	a
4	b	34	a
5	d	35	a
6	c	36	d
7	b	37	c
8	a	38	b
9	b	39	b
10	d	40	c
11	c	41	d
12	b	42	a
13	d	43	b
14	a	44	c
15	a	45	d
16	b	46	c
17	b	47	b
18	b	48	a
19	b	49	c
20	a	50	b
21	d	51	c
22	d	52	a
23	b	53	d
24	b	54	a
25	b	55	d
26	d	56	b
27	c	57	c
28	c	58	d
29	d	59	a
30	d	60	c

More information

More information

Notes

Further reading

Required reading

BS 7671: 2008 Requirements for Electrical Installations, IEE Wiring Regulations Seventeenth Edition, published by the IEE, London, 2008

Additional reading

On-Site Guide: BS 7671: 2008, published by the IEE, London, 2008

The Electrician's Guide to Good Electrical Practice, published by Amicus, 2005

Electrician's Guide to the Building Regulations, published by the IEE, London, 2005

IEE Guidance Notes, a series of guidance notes, each of which enlarges upon and amplifies the particular requirements of a part of the Wiring Regulations, Seventeenth Edition, published by the IEE, London:
– Guidance Note 1: *Selection and Erection of Equipment*, 5th Edition 2008
– Guidance Note 2: *Isolation and Switching*, 5th Edition 2008
– Guidance Note 3: *Inspection and Testing*, 5th Edition 2008
– Guidance Note 4: *Protection Against Fire*, 5th Edition 2008
– Guidance Note 5: *Protection Against Electric Shock*, 5th Edition 2008
– Guidance Note 6: *Protection Against Overcurrent*, 5th Edition 2008
– Guidance Note 7: *Special Locations*, 3rd Edition 2008

Brian Scaddan, *Electrical Installation Work*, published by Newnes (an imprint of Butterworth-Heinemann), 2002

John Whitfield, *Electrical Craft Principles*, published by the IEE, London, 1995

Online resources

City & Guilds www.cityandguilds.com
The City & Guilds website can give you more information about studying for further professional and vocational qualifications to advance your personal or career development, as well as locations of centres that provide the courses.

Institution of Engineering and Technology (IET) www.theiet.org
The Institution of Engineering and Technology was formed by the amalgamation of the Institution of Electrical Engineers (IEE) and the Institution of Incorporated Engineers (IIE). It is the largest professional engineering society in Europe and the second largest of its type in the world. The Institution produces the *IEE Wiring Regulations* and a range of supporting material and courses.

SmartScreen www.smartscreen.co.uk
City & Guilds' dedicated online support portal SmartScreen provides learner and tutor support for over 100 City & Guilds qualifications. It helps engage learners in the excitement of learning and enables tutors to free up more time to do what they love the most – teach!

BRE Certification Ltd www.partp.co.uk

British Standards Institution www.bsi-global.com

CORGI Services Ltd www.corgi-gas-safety.com

ELECSA Ltd www.elecsa.org.uk

Electrical Contractors' Association (ECA) www.eca.co.uk

Joint Industry Board for the Electrical Contracting Industry (JIB)
www.jib.org.uk

NAPIT Certification Services Ltd www.napit.org.uk

National Inspection Council for Electrical Installation Contracting (NICEIC) www.niceic.org.uk

Oil Firing Technical Association for the Petroleum Industry (OFTEC)
www.oftec.co.uk

Notes

Further courses

City & Guilds Level 3 Certificate in Inspection, Testing and Certification of Electrical Installations (2391-10)

This course is aimed at those with practical experience of inspection and testing of LV electrical installations, who require to become certificated possibly for NICEIC purposes. It is not suitable for beginners. In addition to relevant practical experience, candidates must possess a good working knowledge of the requirements of BS 7671 to City & Guilds Level 3 certificate standard or equivalent.

City & Guilds Level 3 Certificate in the Code of Practice for In-service Inspection and Testing of Electrical Equipment (2377)

This course, commonly known as PAT/Portable Appliance Testing, is for staff undertaking and recording inspection and testing of electrical equipment. The course includes a practical exercise. Topics covered: equipment construction, inspection and recording, combined inspection and testing, and equipment.

City & Guilds Level 2 Certificate in Fundamental Inspection, Testing and Initial Verification (2392-10)

This qualification was developed to meet industry needs and to provide candidates with an introduction to the initial verification of electrical installations. It is aimed at practising electricians who have not carried out inspection and testing since qualifying, those who require update training and those with limited experience of inspection and testing. Together with suitable on-site experience, it would also prepare candidates to go on to the Level 3 Certificate in Inspection, Testing and Certification of Electrical Installations (2392-20).

City & Guilds Building Regulations for Electrical Safety

This new suite of qualifications is for Competent Persons in Domestic Electrical Installations (Part P of the Building Regulations). The qualifications consist of components for specialised domestic building regulations and domestic wiring regulations routes as well as a component for Qualified Supervisors.

JIB Electrotechnical Certification Scheme (ECS) Health and Safety Assessment

This Health and Safety Assessment is a requirement for electricians wishing to work on larger construction projects and sites in the UK and the exam is an online type very similar in format to the GOLA tests. It is now a mandatory requirement for holding an ECS card, and is a requirement for all members of the ECS. Please refer to www.jib.org.uk/ecs2.htm for details.

Notes

Notes